A. T. Fomenko
V. V. Kalashnikov
G. V. Nosovsky

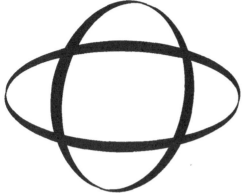

GEOMETRICAL
and STATISTICAL METHODS
of ANALYSIS of STAR
CONFIGURATIONS

DATING PTOLEMY'S ALMAGEST

CRC Press
Boca Raton Ann Arbor London Tokyo

First published 1993 by CRC Press
Taylor & Francis Group
6000 Broken Sound Parkway NW, Suite 300
Boca Raton, FL 33487-2742

Reissued 2018 by CRC Press

Library of Congress Cataloging-in-Publication Data

Fomenko, A. T.
    Geometrical and statistical methods of analysis of star configurations: Dating Ptolemy's
Almagest /
        A. T. Fomenko, V. V. Kalashnikov, G. V. Nosovsky.
        p.   cm.
    Includes bibliographical references and index.
    ISBN 0-8493-4483-2
    1. Stars—Catalogs—Methodology. 2. Ptolemy, 2nd cent. Almagest. I. Kalashnikov, Vladimir
Viacheslavovich. II. Nosovsky, G. V. III. Title.
QB65.F65   1993
523.8'0212—dc20   93-12834

A Library of Congress record exists under LC control number: 93012834

Publisher's Note
The publisher has gone to great lengths to ensure the quality of this reprint but points out that some imperfections in the original copies may be apparent.

Disclaimer
The publisher has made every effort to trace copyright holders and welcomes correspondence from those they have been unable to contact.

ISBN 13: 978-1-315-89315-0 (hbk)
ISBN 13: 978-1-351-07225-0 (ebk)

Visit the Taylor & Francis Web site at http://www.taylorandfrancis.com and the
CRC Press Web site at http://www.crcpress.com

Contents

the "Ptolemaic chronology", that is, Ptolemy's concepts of global chronology (nowadays concealed by the erroneous tradition of recalculating Ptolemy's dates into the years AD). It turns out that similar concepts can be found in several sources of the 13th–14th centuries. Thus, the *Almagest* keeps to a chronological tradition, nowadays forgotten, but actual in the 13th–14th centuries, which differs much from the chronology we are used to today.

The book is concluded by the *Addendum*, containing a brief review of problems connected with dating the *Almagest* as a whole. We treat this material as supplementary, and do not use it in the main body of the book, although it is probably of some epistemological interest.

The book is supplemented with tables containing some astronomic data we use in the text.

The book contains a lot of material represented in tabular and graphical form. We call reader's attention to figures and graphs, which contain much important information, necessary for a fuller understanding of the book. We number the figures and tables consecutively within every chapter and chapter number precedes the number of the figure. Thus "see Figure 2.1" means "see Figure 1 in Chapter 2".

We use the techniques of mathematical statistics, modern geometry, celestial mechanics and astrometry. Therefore some chapters require an acquaintance with basic mathematical notions. Yet, we tried to make the mathematics we used as simple as possible, and we hope that this book will be accessible to a reader familiar with the elements of mathematics at the level of a second-year student in mathematics. The book is intended not only for specialists in natural sciences, but also for the historians interested in modern mathematical and statistical methods. See also the book: A. T. Fomenko, *Empirico-Statistical Methods for Analysis of Narrative and Numerical Sources with Applications to the Problems of Ancient and Medieval History and Chronology*, vols. 1,2. Kluwer Acad. Publ. (in print).

The authors are indebted to Academicians E. P. Velikhov, Yu. V. Prokhorov, Yu. I. Zhuravlev, B. V. Gnedenko and A. S. Zaimovski, Professors V. M. Zolotarev, V. M. Kruglov, V. V. Kozlov, V. K. Abalakin, V. G. Demin, A. V. Nagaev, Yu. N. Tyurin, Yu. K. Belyaev, I. G. Zhurbenko, E. V. Chepurin, Yu. M. Sukhov and S. A. Aivazian for helpful discussions and support they provided in writing this book. We thank Professors H.-J. Lenz, T. Z. Nguen, Yu. V. Deikalo, E. S. Gavrilenko, M. R. Vovchenko, V. V. Kalashnikov (junior), A. A. Borisenko, Yu. G. Fomin, C. Yu. Zholkov, T. S. Turova, O. Yu. Soboleva and Yu. A. Tyurina for their valuable help in processing numerical data, analyzing sources (rare printed editions and manuscripts) and helpful consultations on the subjects.

We thank E. K. Orlova for her selfless help in preparation of the manuscript of this book.

A. T. Fomenko
V. V. Kalashnikov
G. V. Nosovsky

Addendum is intended for a reader wishing to proceed with the study of the questions we raise in the main body of the book toward understanding the origins of the data. A reader interested in mathematical and astronomical aspects alone may confine himself to the main body of the book.

The structure of the book is the following.

The *Introduction* provides a brief review of the contents of the *Almagest*, and in particular of its star catalog. We also give a brief review of other star catalogs and explain our interest in the problem of dating catalogs.

Chapter 1 provides some necessary information from astronomy, astrometry and history of observational equipment and methods for measuring coordinates of stars.

In *Chapter 2*, we carry out a preliminary analysis of the star catalog of the *Almagest*. We discuss here various problems that arise in connection with the catalog (for example, the ambiguity in identification of stars), the accuracy of altitudes and longitudes in the catalog, and some peculiarities of the catalog (such as the Peters' sine curve).

In *Chapter 3*, we analyze some attempts to date the star catalog of the *Almagest* based on the most obvious ideas. We show that no straightforward elementary methods lead to a reliable date, and reveal the difficulties behind these failures.

In *Chapter 4*, we start the description of our new method for dating star catalogs. Here we discuss the "Who is who?" problem, the problem of identification of the stars described in the catalog with the ones known in modern astronomy.

Chapter 5 presents mathematical backgrounds for the statistical analysis of the catalog. Here we classify various errors that occur in the catalog, and suggest methods for their detection and for compensation for the systematic component.

In *Chapter 6*, we carry out a global statistical processing of the catalog and of its basic parts. We apply several statistical characteristics to various pieces of the celestial sphere, which enables us to distinguish the "well-measured" and "poorly measured" pieces. The ensuing decomposition of the sky into the "homogeneous areas" (with contrasting accuracy of measurement) implies a new view of the structure of the *Almagest*.

In *Chapter 7*, we apply two different dating procedures, statistical and geometrical, to the catalog of the *Almagest*; the two estimates turn out to agree.

In *Chapter 8*, we suggest an explanation for the "Peters' sine curve", based on the previous results; we also discuss here the value of the angle between the equatorial plane and the ecliptic given in the *Almagest*.

In *Chapter 9*, we apply our method to the catalogs of Tycho Brahe, Ulugh Beg, Hevelius and Al Sûfi (As-Sûfi).

Chapter 10 is devoted to determination of the date using other parts of the *Almagest*. The ensuing results demonstrate perfect agreement with our date for the star catalog. Finally, we obtain the period of time that captures the observations fixed in the *Almagest* (500–1350 AD), and reconstruct

R. Newton by N. A. Morozov in his fundamental book *History in the Light of Natural Sciences*, published in 1928–1932 under the title *Khristos* (*Christ*) (see Ref. 4). It should be noted that the astronomical and mathematical arguments of N. A. Morozov are diverse from the ones of R. Newton, but they lead to a similar conclusion about the necessity of a revision of the traditional views of the *Almagest*. A lot of additional criticism on the subject can be found in the cycle of works of A. T. Fomenko[5-13], devoted to the development of new empirico-statistical methods for detecting dependent narrative texts and for dating the events they describe (in particular, astronomic events).

We stress, however, that the investigations we expose in this book are completely independent of the methods and arguments used in the afore-mentioned works and that we do not use the hypotheses suggested therein.

In this book we suggest a new method for dating ancient star catalogs. The method uses, in particular, the investigation of proper motions of stars. Since these motions are now measured with a very high accuracy (on the basis of astronomic observations of the last two centuries), it is possible to compute the positions of stars in the past. Comparing these with the ones indicated in a star catalog, we can try to determine the time when the observations were made, and consequently the approximate time of compilation of the catalog. However, a practical implementation of this seemingly simple idea encounters major difficulties, both of technical and fundamental nature. Coping with these difficulties requires the new statistico-geometrical method we present in this book. The foundations of the method have been exposed in Refs. 14 and 15. Our approach involves both statistical and geometrical ideas; the latter are necessary because of the geometrical nature of the object we deal with, the evolution of a point set (the set of stars) in the celestial sphere.

We have tested the method on some reliably dated medieval star catalogs, and also on some artificially created catalogs. In the latter case the catalogs were compiled with the help of a computer; of course, the compiler knew the "date of compilation", but the researcher did not. The date was sealed in an envelope to be unsealed only after getting a date from the method. The procedure proved the efficiency of the method: the "date of compilation" was always within the interval it produced.

Then we applied the method to the star catalog of the *Almagest*. The results thus obtained contradict the traditionally accepted date and imply the necessity of its considerable "rejuvenating".

The main body of this book does not involve any historical questions or questions concerning the origins of the data. Thus, we concentrate on the contents of the star catalog itself, and do not even raise any questions concerning the rest of the *Almagest* (the star catalog constitutes the seventh and the eighth books of the *Almagest*).

However, for the reader's convenience, we have supplemented the book with the *Addendum* containing an exposition of some problems and conjectures on dating the *Almagest* as a whole. We should stress once more that the main body of the book is entirely independent of the *Addendum*. The

Preface

This book is devoted to a problem that lies at the crossroad of several sciences: statistics, geometry, celestial mechanics and computational astronomy, the problem of dating ancient star catalogs from an analysis of their contents, on the basis of modern knowledge of how the visible picture of the sky evolves with time. A vivid example is the problem of dating the star catalog of the famous Ptolemy's *Almagest*. The problem has a long and involved history; see a review of publications on the subject in the book of R. Newton[1].

The *Almagest* is traditionally attributed to Claudius Ptolemy (about the 2nd century AD). Yet, some investigations (mainly, the ones carried out in the 18th–19th centuries) revealed some contradictions between the astronomic data contained in the catalog and the astronomic reality of the 2nd century AD. This led to a hypothesis that Ptolemy had in fact used for the *Almagest* a star catalog compiled by Hipparchus (whose lifetime is traditionally attributed to the 2nd century BC), presumably having added some observations of his own. The reader can find a discussion of this hypothesis (and some others) in classical works[2,3]. A more recent book of R. Newton[1] presents a thorough statistical and astronomical analysis of the *Almagest* as a whole, and in particular of the star catalog it contains. R. Newton contends that his analysis gives an irrefutable proof of most observational data contained in the catalog being counterfeit. In any case R. Newton insists on the necessity of an overall revision of our views of the position and the role of the *Almagest* in the history of science. In fact, a similar conclusion and the inference that an essential redating of the *Almagest* is necessary had been suggested long before

Introduction

1. Brief description of the *Almagest*

The *Almagest* is a famous work of an Alexandrite astronomer, mathematician and philosopher Claudius Ptolemy, whose lifetime is traditionally attributed to the 2nd century AD. We give some information about Ptolemy below; it should be noted, however, that "history treated somewhat strangely the person and the works of Ptolemy. Historians of his time never mention his life and activities . . . . No facts of his life, neither the dates of his birth and death are known" (Ref. 16, p. 96).

It is traditionally considered that the *Almagest* was created in the reign of Roman emperor Antoninus Pius (131–161 AD).

The *Almagest* contains 13 books, about 1000 pages in total volume (in modern editions).

The first book contains basic concepts and constructions, of which the following should be mentioned: 1) The firmament is spherical, and rotates like a sphere; 2) The earth is a sphere, disposed at the center of the universe; 3) The earth can be considered as a point in comparison with the distances to the sphere of fixed stars; 4) The earth does not alter its position in space (does not move). As Ptolemy notes, these principles are based on the conclusions of Aristotle's philosophy. Further, the first and the second books contain an exposition of elements of spherical astronomy (theorems on spherical triangles, a method for calculating arcs (angles) from the lengths of their spans, etc.).

1

The third book presents a theory of visible solar motion and a discussion of the dates of equinoxes, the length of the year, etc. The fourth book treats the length of the synodic month and the theory of lunar motion. The fifth book is devoted to the construction of some astronomic instruments and to a further development of the theory of the moon. The sixth book exposes a theory of solar and lunar eclipses.

The famous star catalog (comprising more than 1000 stars) is contained in the seventh and eighth books of the *Almagest*. The books contain the catalog and a discussion of properties of fixed stars, of motion of the celestial sphere, etc.

The last five books of the *Almagest* are devoted to the theory of motion of planets (Ptolemy considers five planets, Saturn, Jupiter, Mars, Venus and Mercury).

2. A brief review of the history of the *Almagest*

It is commonly accepted that the *Almagest* was created in the reign of Antoninus Pius (131–161 AD) and that the last observation included therein had been made on February 2, 141 AD. The Greek title of the *Almagest*, μαθηματικη συνταξις, implies that the *Almagest* exposes the state-of-the-art of contemporary Greek astronomy. Nowadays it is not known whether any other astronomical treatises comparable to the *Almagest* existed at the time. Usually, the tremendous success of the *Almagest* (with astronomers, as well as with other scientists) is attributed[17] to the loss of most of the astronomic treatises of the time. The *Almagest* had become the basic textbook in astronomy (as is considered nowadays) for more than a thousand years. It influenced greatly the late medieval astronomy, both in Islamic and Christian regions, up to the 16th century. The influence of this book might be only compared to the influence of Euclid's *Elements* on the medieval science.

As noted, for example, by Toomer (Ref. 17, p. 2), it is extremely difficult to trace the history of the *Almagest* and its influences from the 2nd century AD to the Middle Ages. Commentaries of Pappus and Theon of Alexandria are the usual source for judgment on the role of the *Almagest* as a standard textbook in astronomy for "advanced students" in the schools at Alexandria in late antiquity. Further, a "period of darkness" comes. We will only note here the following description of this period: "After the exciting blossoming forth of antique culture, on the European continent a long period of stagnation, sometimes even of regress, began, usually referred to as Middle Ages . . . Over more than 1000 years not a single essential discovery in astronomy was made . . . " (Ref. 17, p. 73).

Furthermore, it is believed that in the 8th and the 9th centuries, in connection with growing interest in Greek science in the Islamic world, the *Almagest* was "raised from the darkness" and was translated, first into Syrian and later, several times, into Arabic. By the middle of the 12th century at least five versions of the translations existed. It is presumed that while in the East (in particular, in Byzantium), the work of Ptolemy, originally written in Greek, was being copied and, to some extent, studied, "all knowledge of it was lost to western Europe by the early middle ages. Although translations from the Greek text into Latin were made in medieval times, the principal channel for the recovery of the *Almagest* in the west was the translation from the Arabic by Gerard of Cremona, made at Toledo and completed in 1175. Manuscripts of the Greek text (of the *Almagest* — *Authors*) began to reach the west in the fifteenth century, but it was Gerard's text which underlay (often at several removes) books on astronomy as late as the Peurbach-Regiomontanus epitome of the *Almagest* ... It was also the version in which the *Almagest* was first printed (Venice, 1515). The sixteenth century saw the wide dissemination of the Greek text (printed at Basel by Hervagius, 1538), and also the obsolence of Ptolemy's astronomical system, brought about not so much by the work of Copernicus (which in form and concepts is still dominated by the *Almagest*) as by that of Brahe and Kepler" (Ref. 17, pp. 2–3).

3. Basic medieval star catalogs

The catalog of the Almagest is the only extant antique star catalog; it is traditionally dated about the 2nd century AD. It is considered, however, that Ptolemy used the star catalog compiled by his predecessor Hipparchus about the 2nd century BC. The *Almagest* catalog (as well as other catalogs of later origin) comprises about 1000 stars, whose positions are described in terms of their longitudes and latitudes (see below for details). After Ptolemy, the "period of darkness and regress" in the history of astronomy (and in the history of all natural sciences) begins, and we know of no other star catalogs up to the 10th century. Finally, only as late as in the 10th century (according to the traditional chronology) was the first medieval catalog created, the one composed by Arabic astronomer As-Sûfi (Abd Al Rahman Al Sûfi, 903–966) in Baghdad. This catalog has come down to us. The next at our disposal is the Ulugh Beg star catalog (1394–1449, Samarkand). The three catalogs are not very precise: they indicate the coordinates of stars to an accuracy within 10 minutes of arc. The next extant is the famous catalog of Tycho Brahe (1546–1601), the precision of which is an order of magnitude better than that of the three preceding catalogs. Brahe's catalog is the acme of skill reached with the help of medieval methods and instruments for astronomical observations. We stop our enumeration here and do not list the catalogs created after Tycho Brahe (there were many, and they are of no interest to us here).

4. Why dating star catalogs is interesting

Every star catalog comprising about 1000 stars is a result of a lot of observations made by an astronomer (even more likely, by a group of professional observers), which required much effort, thoroughness and professionalism, and also an utmost use of available measuring instruments, which were made at the highest contemporary level. Moreover, a catalog required a proper astronomic theory, a world view. Thus, every ancient catalog is a focus of the astronomic mind of the age. So, analyzing a catalog we can learn much about the available accuracy of measurement, the astronomic ideas of the time, etc.

But to understand properly the results of the analysis, we need to know the time when the catalog was compiled. Any variation in the date alters automatically our estimates and conclusions about the catalog. Meanwhile, to determine the date of compilation of a catalog is far from easy. This is very well seen in the case of the *Almagest*. First (in the 18th century), the traditional version attributing the catalog to Ptolemy, about the 2nd century AD, was indisputably accepted. In the 19th century, a more thorough analysis of longitudes of stars indicated in the *Almagest* showed (we describe the details below) that they are more likely to belong to the 2nd century BC, that is, to the time of Hipparchus.

The catalog, contained in the seventh and the eighth books of the *Almagest*, comprises 1028 stars (three of which are duplicates). It contains not a single star that could be observed by Ptolemy from Alexandria, but not by Hipparchus from Rhodes. Moreover, Ptolemy claimed that he had determined, from comparison of his observations with the ones of Hipparchus and others, the magnitude of precession 38′ (which is erroneous), treated by Hipparchus as the least possible value, and by Ptolemy, as the final estimate. The positions of stars as indicated in Ptolemy's catalog are nearer to their real positions in the time of Hipparchus, with the purported 38′ yearly correction, than to their real positions in the time of Ptolemy. So, it looks very likely that the catalog is not a result of Ptolemy's own observations, but the catalog of Hipparchus, corrected for precession, with a few alterations from observations of Ptolemy or other astronomers (see Ref. 2, pp. 68–69).

Thus, in this case the date of compilation of the catalog acquires a paramount importance. For several centuries astronomers and historians of astronomy analyzed the catalog (and the *Almagest* in the whole) trying to "sort" the data contained therein to separate the observations of Hipparchus from the ones of Ptolemy. A lot of literature is devoted to this dating problem. We do not dwell on a review of this literature here; an interested reader will find a guide thereto in Ref. 1.

In this book we consider the question: Is it possible to create a method for dating star catalogs "intrinsically", that is, using only the numeric information contained in the coordinates of stars indicated in the catalog? Our answer is YES. We have worked out such a method, tested it on several reliably dated catalogs and applied it, in particular, to the *Almagest*. The reader will learn our results from this book.

Part One
Preliminary Analysis

Chapter 1
Some Concepts of Astronomy and History of Astronomy

1. Ecliptic, equator and precession

Let us consider the orbital motion of the earth around the sun. It is conventional to treat this motion as the motion of the so-called *barycenter*, the mass center of the system earth-moon. The barycenter is about six thousand kilometers from the center of the earth (hence under the earth's surface). This distance is unessential for our further treatment, so we will make no distinction between the motion of the earth and the motion of the barycenter. Gravitational pull from other planets brings about steady rotation of the orbital plane of the barycenter. The principal sinusoidal component of the rotation has a very large period, and in small intervals of time may be treated as linear. The real motion is the sum of this component with minor oscillations, which we will neglect. The rotating plane that contains the orbit is called the *ecliptic plane*. The circumference where the ecliptic plane meets the *sphere of fixed stars* is called *the ecliptic*. We assume that the center of the sphere of fixed stars O lies in the ecliptic plane (Figure 1.1). Since the ecliptic moves, it is called the *moving ecliptic*. The position of the ecliptic at a given moment of time is called the *instantaneous ecliptic*. For example, we can speak of the *instantaneous ecliptic of January 1, 1900*. It should be clear that we can use any fixed instantaneous ecliptic as a frame of reference for other ecliptics.

Celestial mechanics usually treats the earth as a *rigid body*. A rotation of a rigid body is usually described in terms of its *moment ellipsoid*, determined by its axes, called the *axes of inertia*. A particular rotation of a rigid body is

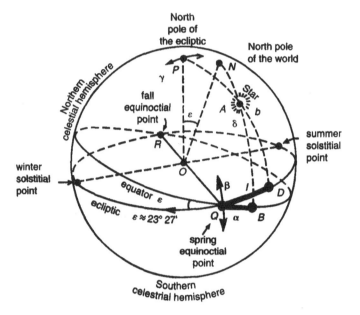

Figure 1.1. The sphere of fixed stars, with the ecliptic and the equatorial coordinate systems.

characterized by the *vector of angular velocity* ω, sometimes called the *instantaneous rotation axis* of the body (in our case, of the earth). Since the axes of inertia A, B, C ($A > B > C$) are orthogonal, we can use them as the axes of a rectangular coordinate system. Now we can consider the projections x, y, z of the vector ω on the axes A, B and C as the coordinates of ω. The rotation of a rigid body can now be described by the *Euler-Poisson equations*:

$$A\dot{x} + (C - B)yz = M_A$$
(1)
$$B\dot{y} + (A - C)xz = M_B$$
$$C\dot{z} + (B - A)xy = M_C$$

where M_A, M_B, M_C are the projections on the axes of a vector M, called the *moment of outer forces about the barycenter*. The moment M is mainly due to the gravitational pull of the sun and the moon on the ellipsoid that is the earth. Usually, the earth is assumed to be an ellipsoid of revolution (that is, the greater semiaxes A and B are assumed equal). The position of M with reference to the axes A, B and C varies with time very fast and in a very complicated way; however, modern theories of lunar and solar motion enable us to compute it to a high accuracy for any moment of time. Consequently, we can solve the Euler-Poisson equations, thus determining the evolution of ω. Usually, the *Tables of the Motion of the Earth on its Axis and around the Sun* by the well-known astronomer S. Newcomb[19] are used to take into account all irregularities of the motion. A study of the Euler-Poisson equations from the point of view of the existence of *exact* solution constitutes an important field of modern theoretical mechanics, physics and geometry; see a short review hereof, for example, in Ref. 20.

The vector of instantaneous angular velocity of the earth ω determines the (instantaneous) *axis of rotation*. The points where the axis of rotation pierces the surface of the earth are called the *instantaneous poles of the earth*, and the points where the axis meets the celestial sphere (the sphere of fixed stars) are called the (North and South) *poles of the world* (Figure 1.1). The intersection of the plane through the center of the earth perpendicular to the axis of rotation with the surface of the earth is called the (instantaneous) *equator*, and its intersection with the celestial sphere is called the (*true*) *equator of the celestial sphere*.

Let us now consider a coordinate system that does not rotate together with the earth, for example, the one associated with the ecliptic. Conventionally, the following axes are used as coordinates in this system: the normal to the ecliptic plane, the axis where the ecliptic plane intersects the equatorial plane (the *equinoctial axis*) and the axis of inertia C. The projections of ω on the three axes are denoted by $\dot{\psi}, \dot{\theta}$ and $\dot{\varphi}$. Thus, we have expanded the velocity of rotation of the earth into three components. What is their geometrical sense? $\dot{\psi}$ is called the *velocity of precession of the earth*. It characterizes the motion of the axis of precession C (the third axis of inertia) along a circular cone about the normal OP (see Figure 1.2); thus, the vector $\omega = ON$ moves along the same cone. Note that the axes ω and OC are very close to each other, so in calculations that do not require high accuracy we may assume that the vector ω is parallel to OC. Because of the precession, the equinoctial axis rotates in the ecliptic plane.

The component $\dot{\theta}$ characterizes variation of the angle θ the axis OC makes with the ecliptic plane. As for $\dot{\varphi}$, it determines the velocity of the earth's rotation about the axis OC; in theoretical mechanics this magnitude is called *the velocity of proper rotation*. This velocity is much greater than $\dot{\psi}$ and $\dot{\theta}$. From the point of view of theoretical mechanics, this reflects the principle according to which a rotation of a rigid body is stable when its axis is close to the axis of the greatest moment of inertia, that is, to the shortest axis of the ellipsoid of inertia.

Thus, $\omega = \dot{\psi} + \dot{\theta} + \dot{\varphi}$ where $+$ stands for summation of vectors. Each of $\dot{\psi}$, $\dot{\theta}$ and $\dot{\varphi}$ is the sum of a constant (or almost constant) component and many minor periodic summands, called *nutations*. Neglecting nutations, we come to the following picture of rotation of the earth.

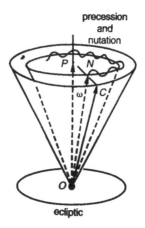

Figure 1.2. Trajectory of motion of the earth's precession.

1) The (almost) constant component of $\dot\psi$ is called the *longitudinal precession*; it moves the axis OC uniformly along a circular cone (see Figure 1.2) at the rate of approximately 50″ per year; the equinoctial axis rotates in the ecliptic plane clockwise if looked at from the North pole of the ecliptic. The vector of precession is directed toward the South pole of the ecliptic.

2) The constant component of $\dot\theta$ is now approximately equal to 0.5″ per year.

3) The constant component of $\dot\varphi$ is the *mean proper rotation of the earth* about the axis OC, anticlockwise if looked at from the North pole of the earth; the period of rotation is 24 hours.

Note that the axis OP (the normal to the ecliptic plane), the vector ω (the instantaneous angular velocity of the earth) and the axis OC lie in the same plane. The precession turns this plane about the axis OP.

Nutational addends in $\dot\psi$, $\dot\theta$ and $\dot\varphi$ distort the above picture of rotation. Therefore the vector ω moves not along an ideal circular cone, but along a "wavy" surface near the cone (Figure 1.2). In Figure 1.2, the trajectory of the endpoint of ω is depicted by a wavy line. Two circumferences in the celestial sphere, the ecliptic and the equator, meet at the angle $\varepsilon \approx 23°27'$ at points Q and R (Figure 1.1). These are the points where the sun passes the equator in its yearly motion along the ecliptic. The point Q, where the sun enters the Northern hemisphere, is called the *spring equinoctial point* (when the sun is at this point, day and night have equal length all over the surface of the earth). The point R is the *fall equinoctial point* (Figure 1.1). As the moving ecliptic turns, *the spring equinoctial point moves steadily along the equator (shifting simultaneously along the ecliptic)*. The rate of this motion of the equinoctial point along the ecliptic is exactly the longitudinal precession. The shift of the equinoctial points thus produces a shift of dates of equinoxes (Figure 1.1).

2. Equatorial and ecliptic coordinates

Recording observations of heavenly bodies requires a convenient coordinate system. Several coordinate systems are used to that end. The *equatorial coordinate system* is defined as follows. Figure 1.1 shows the North pole N and the celestial equator, containing the arc QB. We may assume with sufficiently high accuracy that the plane of the celestial equator contains also the earth's equator; furthermore, we assume that the center of the earth coincides with the center O of the celestial sphere; Q is the spring equinoctial point. Let A be a fixed star and NB the meridian through the North pole and A; here B is the point where the meridian meets the equatorial plane. The arc $QB = \alpha$ is the *equatorial longitude* of the star A, also called *the direct ascent* of the star. The ascent is counted in the direction opposite to the one of the motion of the spring equinoctial point Q. Consequently, due to precession, *the direct ascents of stars slowly increase with time*. The arc δ of the meridian AB in Figure 1.1 is called the *equatorial latitude*, or the *declination* of the star A. If we neglect oscillations of the ecliptic, *the declinations of stars in the Northern hemisphere slowly decrease with time* (because of the shift of the spring equinoctial point), *and the declinations of stars of the Southern hemisphere slowly increase*. The diurnal rotation of the earth does not affect declinations, and the direct ascents vary uniformly at the velocity of the earth's rotation.

Another frequently used system (especially in ancient catalogs) is the *ecliptic coordinate system*. Consider the celestial meridian through the pole of the ecliptic P and the star A (Figure 1.1). The meridian meets the ecliptic plane at a point D. The arc QD in Figure 1.1 is the *ecliptic longitude l*, and the arc AD is the *ecliptic latitude b* of A. Because of precession, the arc QD increases with time (at the rate of about 1° per century), so *the ecliptic longitudes uniformly increase with time*. If we neglect oscillations of the ecliptic, we can assume to a first approximation that the ecliptic latitudes b do not vary with time. This circumstance made the ecliptic coordinates popular among medieval astronomers. The advantage of the ecliptic coordinates over equatorial is for uniform (and easily computable) variation of l and the constancy of b. As for the variations of equatorial coordinates generated by precession, they are described by more complicated formulas (taking into account the turn through the angle between the equator and the ecliptic). This is the reason why medieval astronomers chose to compile their catalogs in ecliptic coordinates, despite the fact that equatorial coordinates are easier to measure from observations. The disappointing discovery of oscillations of the ecliptic brought about the use of equatorial coordinates in modern catalogs.

3. Methods of measuring equatorial and ecliptic coordinates

Here we dwell for a while on a brief description of concrete measurements of equatorial and ecliptic coordinates. We will describe a simple geometrical idea that underlies such measuring instruments as *quadrant, sextant, meridian circle*, etc.

Suppose the vantage point H is at the latitude φ on the surface of the earth (Figures 1.3 and 1.4). It is not difficult to determine the straight line HN' towards the North pole of the world (the line parallel to ON). Find the meridian through H and erect a vertical wall along the meridian (Figures 1.3, 1.4). If we draw on the wall the ray HN', we can also find the equatorial line HK' parallel to OK, lying at a right angle off HN'. Sectoring the right angle into degrees of arc, we get an astronomic goniometer. The idea of this instrument underlies modern meridian instruments. The instrument can be used to measure declinations of stars, i.e., their equatorial latitudes, and to fix the moments when the stars pass the meridian. Since we can determine the equatorial plane (at a given latitude of the vantage point) with a sufficiently high accuracy from a series of consequent independent observations, this instrument enables us to measure declinations with a fairly high accuracy. Meanwhile, as can be seen from the above description of elementary concepts of celestial mechanics, *measuring longitudes requires fixing moments of stars' passing the meridian, for which we need either a sufficiently precise clock, or an additional instrument for fast measurement of longitudinal distance between the star and the meridian.* In any case, measurement of longitudes is a much more complicated operation, so it looks likely that medieval astronomers measured direct ascents with much lower accuracy than declinations.

To determine ecliptic coordinates, the observer H must first determine the position of the ecliptic in the sky. This nontrivial procedure requires a fair knowledge of the geometry of basic elements of motion of the earth and the sun. Some ancient methods for determination of inclination of the ecliptic to the equator and for finding the position of the equinoctial axis are described in Ref. 1. It is important to note that an immediate measurement of ecliptic coordinates of stars is impossible unless we have a clockwork able to compensate for the rotation of the earth and to keep fixed the direction towards the equinoctial point. The obvious difficulty of this problem made the astronomers as they calculated ecliptic coordinates either use the formulas for the turns of the celestial sphere, or celestial globes carrying frames both of equatorial and ecliptic coordinates, thus making it possible to recalculate immediately. Of course, this procedure inevitably added errors originating in determination of the position of the ecliptic in relation to the equator and to the equinoctial axis.

The above brief discussion of methods of measuring ecliptic coordinates leads to the conclusion that the following algorithm was used:

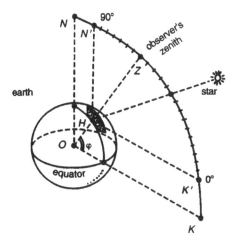

Figure 1.3. The procedure of measurement of equatorial latitude of a star with the help of a meridian circle (1).

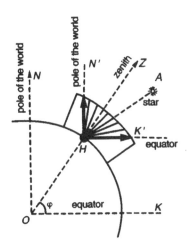

Figure 1.4. The procedure of measurement of equatorial latitude of a star with the help of a meridian circle (2).

1) Find equatorial coordinates (latitudes were determined with a higher accuracy than longitudes).
2) Calculate the position of the ecliptic and the equinoctial axis in relation to the equator.

3) Recalculate equatorial coordinates into ecliptic with the help of trigono-
metric formulas, or an instrument, or a double-framed celestial globe.

Furthermore, since all observational instruments were earthbound, the
above algorithm is the only realistic way of finding ecliptic coordinates the
medieval astronomers could use. The fact that the observational instrument
is attached to the surface of the earth and hence shares the earth's rotation
means that the instrument is bound to the equatorial coordinate system.

Below, we will get a confirmation for the assumption that the above algorithm (or a similar procedure) was used for the star catalog of the *Almagest*
from our statistical analysis.

4. Modern starry sky

1. If we want to date an ancient or medieval star catalog from the coordinates of stars it contains, we must be able to compute positions of stars at
various moments of time in the past. The starting point is the now existing
starry sky. We will only be interested in coordinates of stars, their proper velocities and their *star magnitudes*, that characterize visible brightness (the less
the star magnitude, the brighter is the star). Star magnitudes are indicated in
the most ancient catalogs. In particular, the *Almagest* indicates magnitudes
for all stars it contains. The scale it uses matches in general with the one now
in use, but modern catalogs indicate fractional values of the magnitudes. For
example, Arcturus, which has magnitude 1 in the *Almagest*, has magnitude
0.24 in modern catalogs[21]; Sirius, also having magnitude 1 in the *Almagest*,
has magnitude −1.6 (negative) in modern catalogs. Thus, Sirius is brighter
than Arcturus, while Ptolemy considered them as equally bright. In the Middle Ages, the brightness (star magnitude) was judged by eye. The color of the
star, the brightness of nearby stars and other factors influenced the result. So,
star magnitudes were determined rather roughly. Nowadays star magnitudes
are measured with the help of photometry. A comparison of the *Almagest*'s
star magnitudes with modern precise values shows[22] that the difference usually does not exceed two units. We used the catalog[21], comprising about nine
thousand stars up to the eighth star magnitude. Recall that only sixth to seventh magnitude stars are visible to the unaided eye, and the catalog of the
Almagest, as Ptolemy claims, contains all stars of the visible part of the sky up
to the sixth magnitude. In fact, though, there are many more stars of sixth
and lesser magnitudes in the visible sky than in Ptolemy's catalog. This
is one of the causes of ambiguities that arise in attempts to identify the stars
in the *Almagest* with the stars in modern catalogs (computed back to the
past).

The astronomer of the 17th century I. Bayer suggested to denote stars
in a constellation by Greek letters: the brightest star is denoted by α, the

second in brightness by β, and so on. For example, α Leo is the brightest star in the constellation Leo. Later on, J. Flamsteed (1646–1720) assigned to stars the numbers (in the constellation): the westernmost star acquired number 1, the next to the east number 2, and so on. The Flamsteed number and the Bayer letter are usually written together in the denotation of a star, for example, 32 α Leo. Furthermore, a star can have a proper name. There are comparatively few "named" stars; the names were only given to the stars which had special significance in antique and medieval astronomy. For example, 32 α Leo has the proper name *Regul* (*Regulus*).

We used the following characteristics of stars from Ref. 21:

1) Direct ascent of the star in 1900, denoted by α_{1900} and measured in hours, minutes and seconds.

2) Declination of the star in 1900, denoted by δ_{1900} and measured in degrees, minutes and seconds of arc.

3) Velocities of the proper motion of the star in declination and in ascent, that is, the projections of the velocity of proper motion on the equatorial coordinate axes in 1900.

The velocities of proper motions of stars are rather small; as a rule, they do not exceed 1″ per year, and the fastest stars visible by unaided eye (o^2 Eri, μ Cas) move at the rate of about 4″ per year. In the interval of time we are interested in, about two to three thousand years long, the proper motion may be assumed uniform in each coordinate in a fixed coordinate system. For us, this coordinate system is the equatorial coordinate system of 1900. For the reader's convenience, we adduce in the *Appendix* two lists of characteristics of stars taken from Ref. 21. Table Ap. 1 is the list of fast stars. It contains all stars whose proper motion in at least one of the coordinates α_{1900}, δ_{1900} is not less than 0.5″ per year. Table Ap. 2 is the list of named stars. The two tables have a common part: some named stars have a notable proper motion; such stars are especially useful for dating purposes (see below).

5. Computation of the starry sky to the past. Catalogs $K(t)$. Newcomb's theory

1. Having at our disposal the coordinates and the velocities of proper motion of stars in our time, we can calculate a precise catalog for an arbitrary epoch. We had to do this many times and for various epochs as we investigated the *Almagest* and other ancient catalogs. Compiling these "theoretical" catalogs, we first computed the positions of stars at the year t in coordinates α_{1900} and δ_{1900}, and then recalculated into ecliptic coordinates l_t and b_t for the year t. Below, we give the necessary formulas making it possible to take into account the precession and, in particular, to recalculate from α_s, δ_s into

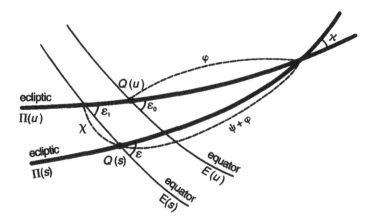

Figure 1.5. Relations between ecliptic and equatorial coordinates in various epochs.

l_u, b_u for any two epochs s and u. These formulas, as well as Figure 1.5 are taken from Ref. 23. They were obtained on the basis of a theory of Newcomb, modified by Kinoshita. The procedure of recalculating coordinates is described in Subsection 2 below. In the formulas, we assume that the epochs u and s are counted in Julian centuries from 2000 AD, and that $\theta = u - s$ (see Figure 1.5).

(1)
$$\varphi(s, u) = 174°52'27''.66 + 3289''.80023u + 0''.576264u^2$$
$$- (870''.63478 + 0''.554988u)\theta + 0''.0245780\theta^2$$

(2)
$$\varkappa(s, u) = (47''.0036 - 0''.06639u + 0''.000569u^2)\theta$$
$$+ (-0''.03320 + 0''.000569u)\theta^2 + 0''.0000500\theta^3$$

(3)
$$\varepsilon(s, u) = 23°26'21''.47 - 46''.81559u$$
$$- 0''.000412u^2 + 0''.00183u^3$$
$$+ (-46''.8156 - 0''.00082u + 0''.005489u^2)\theta$$
$$+ (-0''.00041 + 0''.005490u)\theta^2 + 0''.0018300\theta^3$$

(4)
$$\varepsilon_0(s, u) = 23°26'21''.47 - 46''.81559u$$
$$- 0''.000412u^2 + 0''.00183u^3$$

$$\varepsilon_1(s, u) = 23°26'21''.47 - 46''.81559u$$

(5)
$$- 0''.000412u^2 + 0''.00183u^3$$

$$+ (0''.05130 - 0''.009203u)\theta^2 - 0''.007734\theta^3$$

$$\psi(s, u) = (5038''.7802 + 0''.49254u - 0''.00039u)\theta$$

(6)
$$+ (-1''.05331 - 0''.001513u)\theta^2 - 0''.001530\theta$$

$$\chi(s, u) = (10''.5567 - 1''.88692u - 0''.000144u)\theta$$

(7)
$$+ (-2''.38191 - 0''.001554u)\theta^2 - 0''.001661\theta^3$$

$$\Psi(s, u) = (5029''.0946 + 2''.22280u + 0''.000264u^2)\theta$$

(8)
$$+ (1''.13157 + 0''.000212u)\theta^2 + 0''.0001020\theta^3$$

We should note that the distinctions between the original Newcomb theory and its modification by Kinoshita[23], which we use here, are unimportant for us: for any moment of time t in the interval we are interested in (600 BC–1900 AD), the difference between the ecliptic coordinates computed from the Newcomb theory and the ones from the modification is negligible in comparison with the errors of the *Almagest*. We used Ref. 23 because the formulas for precession are given there in a form convenient for computer calculations.

2. Let us now describe in details the algorithm of compilation of the catalog $K(t)$ reflecting, according to Newcomb's theory, the sky at the moment t. Henceforth we consider t to be an arbitrary moment in the interval 600 AD–1900 BC, counted back in Julian centuries from 1900; thus, for example, $t = 1$ corresponds to 1800 AD, $t = 10$ to 900 AD, and $t = 18$ to 100 AD (the several days' difference that accumulates because of the difference between Julian and Gregorian calendars is absolutely immaterial for our purposes). The reason for this somewhat strange denotation is its matching the existing computer programs and our wish to avoid confusion that could initiate a change of notation. We will compare the catalogs $K(t)$ for various values of t with the ancient catalog we study (say, with the *Almagest*); t will serve as an a priori date for the catalog. Therefore $K(t)$ are to be compiled in ecliptic coordinates of the epoch t, because as we have already noted, ancient and medieval star catalogs used these coordinates.

So, suppose a star has equatorial coordinates $\alpha^0 = \alpha^0_{1900}$ and $\delta^0 = \delta^0_{1900}$ in a modern star catalog (say, in Ref. 21). These coordinates show the position of the star in 1900 in the spherical coordinate system the equator of which coincides with the earth's equator (hence lies in the plane of earth's rotation, which, as we have noted above, changes with time) in 1900. We need to determine the coordinates l_t and b_t (that is, coordinates in the spherical coordinate

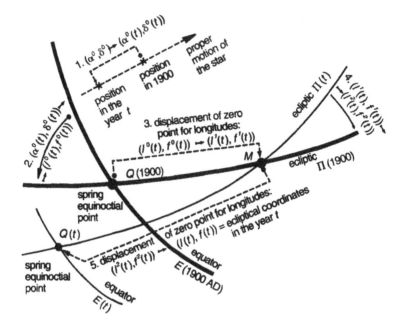

Figure 1.6. The recalculation of ecliptic coordinates of January 1, 1990 into ecliptic coordinates of an arbitrary epoch, taking into account the proper motion of stars.

system whose equator is the ecliptic in the year t). To that end it suffices to do the following (see Figure 1.6):

1) *Find the coordinates of the star* $\alpha^0(t)$ *and* $\delta^0(t)$ *for the year* t *in the equatorial coordinates of 1900.* This can be done with the help of the proper motion velocities v_α and v_δ in the coordinates α and δ (see the fifth and the sixth columns of Tables Ap. 1 and Ap. 2). We have:

$$\text{(9)} \qquad\qquad \alpha^0(t) = \alpha^0_{1900}(t) = \alpha^0 - v_\alpha t$$

$$\text{(10)} \qquad\qquad \delta^0(t) = \delta^0_{1900}(t) = \delta^0 - v_\delta t$$

Indeed, as we have noted above, within the interval of time we are interested in, the proper motion of stars may be treated as uniform. The minuses in (9) and (10) come from our counting time to the past, while the signs of v_α and v_δ correspond to the natural time count.

2) *Pass from coordinates* α_{1900}, δ_{1900} *to the coordinates* l_{1900}, b_{1900}. This gives us coordinates $l^0(t)$ and $b^0(t)$ of the star in the year t in the spherical coordinate system bound to the ecliptic of 1900.

We have:

(11) $\qquad \sin b^0(t) = -\sin \alpha^0(t) \cos \delta^0(t) \sin \varepsilon^0 + \sin \delta^0(t) \cos \varepsilon^0$

(12) $\qquad \tan l^0(t) = \dfrac{\sin \alpha^0(t) \cos \delta^0(t) \cos \varepsilon^0 + \sin \delta^0(t) \sin \varepsilon^0}{\cos \alpha^0(t) \cos \delta^0(t)}$

where

(13) $\qquad\qquad\qquad\qquad \varepsilon^0 = 23°27'8''.26$

These formulas enable us to restore uniquely the values of $b^0(t)$ and $l^0(t)$, because $-90° < \alpha < 90°$ and $|l^0(t) - \alpha^0(t)| < 90°$. The angle ε^0 is the inclination of the ecliptic to the equator in 1900 (see (4), where we put $u = -1$ to pass from year 2000 to 1900).

3) *Pass from coordinates* l_{1900} *and* b_{1900} *to the coordinates* l^1 *and* b^1, which are also bound to the ecliptic of 1900, but whose zero point is at the intersection of the ecliptic of 1900 $\Pi(1900)$ and the ecliptic of the year t $\Pi(t)$. The two coordinate systems are connected by the relations

$$l^1(t) = l^0(t) - \phi$$

(14) $\qquad\quad b^1(t) = b^0(t)$

$$\phi = 173°57'38''.436 + 870''.0798t + 0''.024578t^2$$

Here ϕ is the arc of $\Pi(1900)$ between the spring equinoctial point of 1900 and the point of intersection of $\Pi(1900)$ and $\Pi(t)$; it can be found from (1) by putting $u = -1$ (then $\Pi(u)$ in Figure 1.5 will correspond to $\Pi(1900)$) and $\theta = -t$. Then $\Pi(s)$ in Figure 1.5 will depict the ecliptic of the epoch t. Indeed, t is counted in centuries from 1900 to the past, and $\theta = s - u$ is counted in centuries from u to the future; since we put $u = -1$, which corresponds to 1900 $(2000 - 100 = 1900)$, we have to put $\theta = -t$ to make the epoch $s = u + \theta$ in (1) correspond to the epoch t.

4) *Pass from the coordinates* l^1 *and* b^1 *to the coordinates* l^2 *and* b^2, the spherical coordinates bound with the ecliptic $\Pi(t)$ and differing from the ecliptic coordinates l_t and b_t only for the choice of zero point of the longitudes. In the coordinates l^2 and b^2, the zero point is the intersection of $\Pi(1900)$ and $\Pi(t)$. The transfer formulas from (l^1, b^1) to (l_t, b_t) are similar to (14); we only have to replace ε^0 by the angle ε_1 between $\Pi(1900)$ and $\Pi(t)$; we have

(15) $\qquad\qquad \varepsilon = -47''.0706t - 0''.033769t^2 - 0''.000050t^3$

This expression can be obtained from (2) by putting $u = -1$ and $\theta = -t$.

5) Finally, we are to *pass from l^2 and b^2 to the ecliptic coordinates l_t and b_t*. This can be done by the formulas

(16)
$$l_t = l^2 + \phi + \Psi$$
$$b_t = b^2$$

where ϕ is as in (10) and Ψ can be obtained from (8) by putting $u = -1$ and $\theta = -t$, that is,

(17) $$\Psi = -5026.''872t + 1.''1314t^2 + 0.''0001t^3$$

The sequence of steps 1)–5) is illustrated in Figure 1.6.

6. Astrometry. Some medieval astronomic instruments

In Section 3 we have exposed a general idea of an astronomic goniometer; an important feature of it is the possibility of a sufficiently accurate determination of the line of the celestial equator. The ray HK' along which the observer's eye is directed does not leave the equator in the process of diurnal rotation. Of course, the setting of the ray HK' depends on the geographical latitude of the vantage point. In principle, one can imagine the plane HLM as attached to the quadrant (Figure 1.7). This plane is parallel to the equatorial plane, and intersects the celestial sphere along the celestial equator. This is in no way affected by the fact that HLM actually does not pass through the center of the earth. Thus, at any point of the earth's surface it is possible to build a stationary instrument (oriented along the meridian) that allows a practically visual observation of the equator. *This makes a reliable measurement of equatorial latitudes of stars possible* (Figure 1.7), for example, at the moment when the star passes the vertical plane of the quadrant. As we already noted, for a professional medieval astronomer measuring equatorial latitudes was not a complicated procedure; it only required accuracy and a sufficient time for observations. In particular, we can expect that a thorough observer should not make a big systematic error in declinations of stars.

Let us now look at particular implementations of this idea in medieval astronomic instruments.

The first instrument, the so-called *meridian circle* is described by Ptolemy (Figure 1.8). The device is a flat metallic ring installed vertically on a firm support in the plane of the meridian. The ring is graduated, for example, into 360 degrees. A smaller ring, rotating freely inside the first ring in the same plane was installed (Figure 1.8). Two small metallic plates with arrows pointing to the divisions on the outer ring were attached at two diametrically opposite points of the inner ring (points P in Figure 1.8). The device is

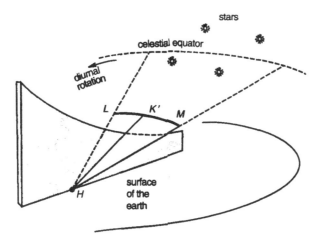

Figure 1.7. Visual determination of the equator's position in the celestial sphere.

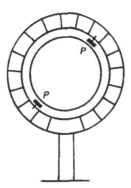

Figure 1.8. Meridian circle.

installed in the meridian plane with the help of a plumb; the direction of the meridian was determined from the shade of a vertical pole at noon. Then the zero division of the outer ring was matched with the zenith. The device could be used for measurement of the altitude of the sun (at the latitude of the vantage point); to that end, the inner ring is turned at noon so that the shade of one of the plates P covered the other. Then the arrow on the upper plate points to the altitude of the sun in degrees on the outer ring. Note that we can read the result *after* fixing the plates; so we can read the altitude after the moment of noon. Furthermore, the meridian circle can be used to determine the angle ε between the ecliptic and the equator.

Another instrument is the *astrolabon* ($\alpha\sigma\tau\rho o\lambda\alpha\beta o\nu$) described by Ptolemy; today this word is translated as *astrolabe*. It should be noted, however, that the meaning of this word altered with time. It is considered that the device we will now describe was used by Ptolemy near the beginning of our era and was called *astrolabon*. But in the Middle Ages, this instrument was already called *armillary sphere*, or *armilla*. Today, some astronomers think (see, for example, Ref. 18) that Ptolemy does not actually describe in the *Almagest* the *astrolabe* (also a medieval term), but does describe the *astrolabon* (or the armillary sphere). R. Newton notes that "Probably, by the second half of the medieval period, it (the term *astrolabe* — *Authors*) had come to mean an instrument for measuring the elevation angle of a celestial object above the horizon. By the time this happened, the kind of instrument just described (following to Ptolemy — *Authors*) was often called an armillary sphere, of which modern telescope mounts are a development" (Ref. 1, p. 145). To avoid terminological confusion, we describe below *two* devices, the *armillary sphere* (Ptolemy's *astrolabon*) and the *astrolabe* (the medieval instrument, the name of which is for some reason practically identical with Ptolemy's *astrolabon*). So, what is the construction of the astrolabon (= armilla)? The basic details are shown in Figure 1.9, and Figure 1.10 exhibits a real medieval armillary sphere. The main detail of the armillary sphere is two metallic perpendicular rings, rigidly fastened together; we will call them the *first ring* and the *second ring* (Figure 1.9). The first ring rotates freely about the axis *NS*, parallel to the earth's axis. The common center of the two rings is the point *O*.

We will now describe how to use the armilla to measure the angle between the ecliptic and the equator. It is best to carry out the measurements on solstice. The corresponding point of the earth's orbit is denoted by *O′* in Figure 1.11 (it makes no difference at the moment whether it is the summer or the winter solstitial point). Consider the plane through the radius vector *SO′* where *S* is the sun, and the earth's axis *NO′*. Since *O′* is the solstitial point, the plane is perpendicular to the ecliptic plane and hence intersects the earth's surface along a meridian (Figure 1.11). Suppose the armilla is at a point of this meridian (we may place the armilla at any point of the earth's surface, but carry out the measurement at noon; at this moment the device is on the meridian where the plane meets the surface). We assume that the observer knows the direction of the earth's axis, and that the axis *NO* is oriented in this direction (parallel to *NO′* in Figure 1.11). Turn the first ring of the armilla around the axis *NS* to place the ring in the plane of the meridian. The ring will be there at the moment when the shade of the outer edge of the ring will cover its inner part. Fixing the plane of the first ring, place the second ring (perpendicular to the first) so that its inner part is in the shade of its outer part. It is clear from Figure 1.11 that then the second ring will lie in the ecliptic plane (more accurately, will be parallel to the ecliptic plane). Furthermore, let $P_1 P_2$ be perpendicular to the second ring. Since both rings are fixed, the line is also fixed, so it determines a pair

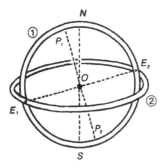

Figure 1.9. Astrolabon (armilla).

Figure 1.10. Medieval armillary sphere.

of points P_1, P_2, on the first ring. Thus the angle P_1ON in Figure 1.11 is well-defined. Clearly, *this angle is equal to the angle between the ecliptic and the equator.* We have described the method the ancient astronomers used. Although the geometrical idea is quite simple, various difficulties introducing errors into the measurement are obvious. In particular, the observer must know (with a sufficient accuracy): a) the direction of the axis ON parallel to the earth's axis, b) the day of solstice, c) the moment of the noon (at the given point of the earth's surface). As R. Newton notes, "The main drawback to this instrument, so far as I can see, is that the rotation of the earth destroys the alignment rather rapidly, so that the instrument must be read quickly" (Ref. 1, p. 145). Indeed, as can be seen from Figure 1.11, the earth's rotation turns the device about the axis NO', making the above considerations incorrect. R. Newton tells that he experimented with a simplified version of the device,

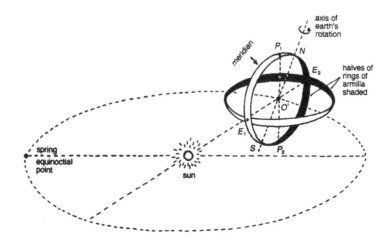

Figure 1.11.   The use of armilla for determination of inclination of the ecliptic on solstice.

and came to the conclusion that placing the second ring into the ecliptic is possible to an accuracy within 2 minutes of arc. If the nearest solstice occurs in no more than a month from the day of observation (for example, if the astronomer made a mistake in determining solstice), then the error in the longitude of the sun must not exceed, in Newton's estimate, 5' (provided that the astronomer can read accurately enough the indication of the latitude ring). Further, R. Newton notes that the overall error probably must not exceed 15'. However, as we have seen, the procedure of measuring the angle of inclination of the ecliptic is fairly delicate, so a 15'–20' error should be treated as natural in using an armilla, assuming that the professional observer was maximally thorough in his observations.

Returning to Figure 1.11, we should note that formally, the points O (the center of the armilla) and O' (the center of the Earth) are different (the distance being equal to the earth's radius), but from the point of view of the above measuring procedure, this distinction is negligible, because the distance is small in comparison with the distance to the sun. Therefore, we can assume $O = O'$ as in Figure 1.11.

Let us now return to measuring ecliptic coordinates with the help of an armilla. Installing the armilla in accordance with the above rules, we get the ecliptic coordinate system. Namely, the plane of the second ring $E_1 E_2$ is parallel to the ecliptic plane, and the points E_1 and E_2 correspond to the solstitial points. Using the divisions on the rings, we can find the points R_1 and R_2 that correspond to the equinoctial points. Thus we get a scale on the second ring $R_1 E_1 R_2 E_2$ with a fixed origin, say at the spring equinoctial point. Consequently, we get an opportunity to measure ecliptic latitudes and

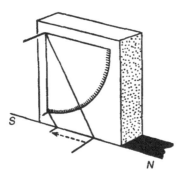

Figure 1.12. Quadrant.

Figure 1.13. Astrolabe.

longitudes of stars. We should recall, however, that the diurnal rotation of the earth upsets alignment of the armilla, so either the measurements are to be carried out as fast as possible, or we have to use a clockwork able to compensate for the rotation and automatically reset the instrument (the latter idea is used in modern observational instruments). To facilitate measuring ecliptic coordinates, one more ring (the third) is added to the armilla, which can rotate about an axis sliding along the ring $R_1 E_1 R_2 E_2$. We will not go into these details, because they are of no importance for us here.

Another instrument is the *quadrant* (Figure 1.12), which can be obtained by installing a spike at the center of a meridian circle (see above), perpendicular to the plane of the circle. The shade of the spike falls on the lower northern part of the ring (see Figure 1.8). The shade can move within a *quarter of the ring*, so it is enough to calibrate only this part of the ring. Thus, the quadrant is a plate with a graduated quarter of the ring, installed in the meridian plane. At noon, the shade of the spike shows the altitude of the sun on the scale.

The fourth instrument is the *astrolabe* (Figure 1.13). The medieval astrolabe is a circle of diameter about half a meter with a fixed calibrated ring at its edge. A plank with sights (*diopters*) is installed on the axis perpendicular to the plane of the circle. The device could be suspended; in the suspended position the disk was directed towards a heavenly body, after which the moving plank was directed towards the body. Thus the altitude of the body above the horizon was determined. For example, this technique could be used for determination of the latitude of the vantage point (from the altitude of the sun at noon). Of course, the measuring could be inexact, because the measuring technique itself is rather rough. It is accepted that this technique could provide the latitude of the vantage point to an accuracy within several minutes of arc.

Chapter 2

Star Catalog
of the *Almagest*.
Preliminary Analysis

1. Structure of the catalog

The star catalog is contained in the seventh and the eighth books of the *Almagest*. In our study we used the canonical edition 1 of the catalog by Peters and Knobel [22] and two complete editions of the *Almagest* [17,24].

We begin with recalling some notation commonly used in history of astronomy.

The catalog is compiled in ecliptic coordinates. The circumference of ecliptic longitudes (from 0° to 360°) is traditionally divided into twelve equal parts, called the *signs of the zodiac* (not to be confused with the zodiacal constellations!). Table 2.1 contains a list of the signs with their traditional denotations and the corresponding longitudinal sectors.

The zodiacal signs are used in compiling catalogs. For example, in the *Almagest* the ecliptic longitude of a star is counted from the beginning of the zodiacal sign the star belongs to. For example, α Ursae Majoris (number 1 in the catalog of the *Almagest*) has longitude Ⅱ 0°10′, which means that it is offset by 10 minutes of arc from the beginning of the zodiacal sign Gemini, disposed at 60° (see Table 2.1). Thus, the absolute ecliptic longitude of the star is 60°10′.

For denoting latitudes, a simpler principle is used in the *Almagest*: the longitudes are counted from the ecliptic (corresponding to zero grade of latitude) to the pole (90 grades of latitude). For example, α Ursae Majoris has

Table 2.1.   Zodiacal signs

Denotation	Sign	Latitudinal interval
♈	Aries	0–30
♉	Taurus	30–60
♊	Gemini	60–90
♋	Cancer	90–120
♌	Leo	120–150
♍	Virgo	150–180
♎	Libra	180–210
♏	Scorpius	210–240
♐	Sagittarius	240–270
♑	Capricornus	270–300
♒	Aquarius	300–330
♓	Pisces	330–360

latitude +66°10′ (in the *Almagest*). The sign "+" or "−" means that the star belongs, respectively, to the Northern or the Southern hemisphere.

Since the signs of zodiac do not match exactly the zodiacal constellations, stars of the same zodiacal constellation may belong to different signs of the zodiac.

The canonical version of the *Almagest* can be found in the work of Peters and Knobel [22].

2. Distribution of well-identified and poorly identified stars of the *Almagest*

Table 6 in Ref. 22 enlists stars of the *Almagest* which are identified with different now known stars by different astronomers. It compares opinions of well-known astronomers: Peters, Baily, Schjellerup, Pierce and Manitius.

We have carried out numeric processing of these data from the following point of view. First, it is useful to display the constellations mentioned in the catalog on the star map. To this end we used a map of the modern sky, exhibiting boundaries of modern constellations. In Figure 2.1, the boundaries are depicted by continuous lines. Of course, this is nothing more but an approximate picture, because the boundaries of ancient constellations were not clearly defined. However, taking into account this roughness, we may assume that Figure 2.1 reflects a qualitative picture of the distribution of constellations of the *Almagest* in the sky. We also used the map adduced in the first editions of the *Almagest* (Latin and Greek editions, the 16th century). Although boundaries of constellations are not clearly delineated in this map (drawn by A. Dürer; the map represents conventional figures of the constellations: Hercules, Pegasus, etc.), a comparison with modern positions of the

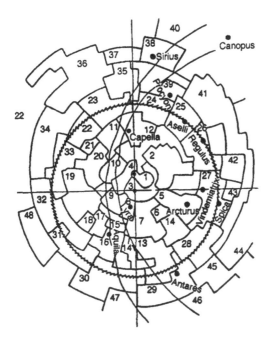

Figure 2.1 (a). Partition of the star map of the *Almagest* into 48 constellations and the twelve named stars (1). (Rims are approximate.) The names of the constellations are shown in Table 2.2.

constellations shows that modern boundaries match the figures represented in the maps of Dürer and of the *Almagest*. Further, the crosshatched circumference in Figure 2.1 depicts the ecliptic, and the broad vertical stripe (tilted to the left) the Milky Way. Of course, here its boundaries are also shown but approximately, indicating the distribution of its most dense regions.

Within the domain corresponding to each constellation, the number of the constellations in the *Almagest* is shown. Their names are given in Table 2.2.

To clarify some terms used below and to give an impression of the source of our arguments, let us describe the structure of Table 6 in Ref. 22, which has already been referred to.

The table consists of six columns. The first column is a list of the numbers of the stars. This enumeration is due to Baily (manuscripts of the *Almagest* do not contain any enumeration). In accordance with this enumeration, the number of the stars is equal to 1028, although there is some divergence between different research about this number because some stars have two entries in the catalog. The list of stars is partitioned into constellations, each having a proper name. The total number of constellations is 48; their names will

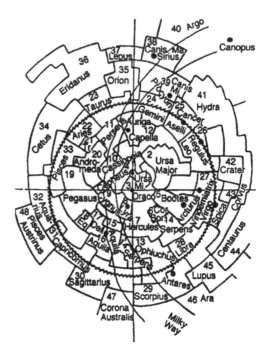

Figure 2.1. (b). Partition of the star map of the *Almagest* into 48 constellations and the twelve named stars (2). (Rims are approximate.)

be listed below. Some constellations are assigned to a group of stars called *informata*. This group contains stars that do not belong to the constellation, whose coordinates are close to coordinates of the stars from the constellation. In other words, the "main" list contains probably the stars included in the skeletons of constellations, and the stars included in the *informata* are a sort of background for "main" stars.

The second column contains a description of stars. Most stars have no names and are endowed with descriptions like "the star on the end of tail," "the star in the middle of the neck," etc. These descriptions are taken from the Latin edition of *Almagest* published in 1528; see Addendum for details.

The third column contains modern names of the stars. Its contents are based on research carried out by various scientists. It is worth mentioning that such identification may have nonunique solutions because of different reasons: fuzziness of the description of stars, moving positions of stars, approximate character of constellations, etc. As we have mentioned, the table in Ref. 22 contains the results of identifications obtained by different astronomers; they do not coincide identically.

The fourth and fifth columns contain ecliptic longitudes and latitudes of the stars, respectively, as has been explained in Section 2.1.

The sixth column contains magnitudes (of the brightness) of the stars.

The *Almagest* contains twelve named stars. Verbal descriptions of these stars always contain the word *vocatur* (named). For example, *vocatur Arcturus* (the star named Arctur). All the twelve stars are depicted in Figure 2.1 by bold black points. These stars are: Arcturus, Regulus, Aselli, Sirius, Procyon, Vindemiatrix, Spica, Lyra (Vega), Capella, Aquila, Canopus, Antares. Most of them either lie in the Milky Way, to the right of the Milky Way, or very close to the Milky Way. Canopus is in fact outside the map, because this star is very much to the South. Further, in Figure 2.1 we display the North pole (in Ursa Minor) and the pole of the ecliptic (in Draco). Let us now look at the order Ptolemy lists the constellations in. To this end we construct a new map, in which instead of the constellations we depict their "centers" as white points; see Figure 2.2. Of course, the center of a constellation can be defined only very roughly, but we are now only interested in a rough qualitative picture. Now let us connect the constellations with stars in the order as in the *Almagest*; see the result in Figure 2.2. The resulting line is a spiral beginning in the Great Bear and unwinding clockwise all along the constellations occurring in the *Almagest*. Let us look more closely at the spiral.

Several pieces can be naturally distinguished in the spiral. First, Ptolemy enlists the constellations with numbers 1 to 8: Ursa Minor, Ursa Major, Draco, Cepheus, Bootes, Corona Borealis, Hercules, Lyra; all disposed in the domain bounded by the zodiacal belt (from the right) and the Milky Way (from the left).

Then Ptolemy lists all constellations which are covered completely by the Milky Way or have a substantial intersection with it: Cygnus, Cassiopeia, Perseus, Auriga, Ophiuchus, Serpens, Sagitta; these constellations have numbers 9 to 15.

Then Ptolemy passes to the domain to the left from the Milky Way (in Figure 2.2) bounded from the left by the zodiacal belt; he lists consequently the constellations Aquila, Delphinus, Equuleus, Pegasus, Andromeda, Triangulum, having numbers 16 to 21.

Then Ptolemy enters zodiacals and makes a round trip about the center of the map, going consequently through all the twelve zodiacal constellations: Aries, Taurus, Gemini, Cancer, Leo, Virgo, Libra, Scorpio, Sagittarius, Capricorn, Aquarius, Pisces. These constellations have numbers 22 to 33.

Then Ptolemy leaves the Northern hemisphere and, crossing the zodiacal belt, goes down to the other hemisphere. On this way, he enlists the constellations Cetus, Orion, Eridanus, Lepus, Canis Major, Canis Minor, Argo, Hydra, Crater, Corvus, Centaurus, Lupus, Ara, Corona Australis, Piscis Austrinus, having numbers 34 to 48, and here the catalog ends.

Thus, the order in which Ptolemy enlists the constellations implies a partition of the star map into several domains.

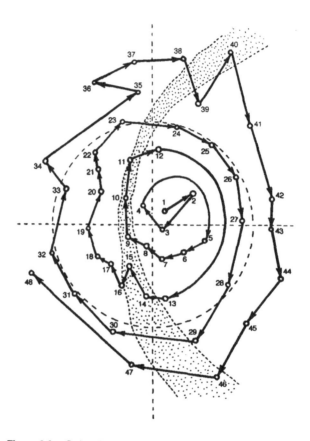

Figure 2.2. Order of constellations in the catalog of the *Almagest*.

Not trying so far to find the reasons for this particular order, let us now outline the domains.

The domain M is the Milky Way, dividing the sky into two parts. Furthermore, *the domain A* is the part of the sky to the right from the Milky Way to the zodiacal belt, including its right rim. The domain contains a domain composed of zodiacal constellations alone, which we will denote by *Zod A.*

Further, *domain B* is the part of the sky to the right from the Milky Way up to the zodiacal belt, including its left rim. We denote the part of the domain *B* consisting of zodiacal constellations by *Zod B.* Finally, *C* is the southern domain of the sky to the left from the zodiac, and *D* is the southern domain to the right from the zodiac (in Figure 2.2).

As we will see below, this partition of the star catalog of the *Almagest* is not accidental, and has some remarkable properties, connected with deep statistical properties of the catalog.

Now we will only note a special (and nontrivial) nature of the order of constellations in the catalog. For example, the compiler could merely list them using a spiral, but passing uniformly from domain A to domain B and back, that is, moving uniformly around the pole. But the order used by Ptolemy is quite different: *he first lists the constellations to the right from M, then the constellations in M, then the ones to the left from M, then the zodiacal constellations, and finally southern stars.* Probably, this order was chosen for some serious reasons. In fact, the reasons are unessential for us, and only the resulting order of constellations is important.

A very important (and not a priori obvious) fact is that the partition of the star catalog is closely connected with the "accuracy characteristics" of descriptions of stars.

As we have already noted, the opinions of specialists concerning identification of several stars of the *Almagest* diverge. Table 6 of Ref. 22 displays the divergences between five most well-known researchers and commentators of the *Almagest*. *The divergence is evidence for the star in question being measured to a too low accuracy to make it possible to find unambiguously its modern position.* Since most stars are not of the first or second magnitude, in order to identify them, one has to take the coordinates indicated in the *Almagest*, then compare them with modern coordinates of stars, and find the star that matches best, that is, disposed most closely to the *Almagest* star. Clearly, this approach (practically, inevitable for non-named, comparatively faint stars) works well only if Ptolemy's coordinates for the star are sufficiently precise. Otherwise, *several possibilities for identification may arise.* The situation becomes especially complicated when the star in question is disposed among several other stars more or less equal in brightness; in this case many possibilities for identification arise, and the final choice is far from being easy. This is the cause of the controversy in identification of stars of the *Almagest*. The "final version" in Ref. 22 may be more or less matching than other versions, and we are not going to go into a detailed discussion of this question, more so because we do not need this for our study. We should hail the accuracy of Peters and Knobel, who conscientiously listed all versions of identification. We will use their Table 6 to carry out some not difficult, but, as we will see, very useful calculations, leading to some conclusions on the accuracy of Ptolemy's measurement in various domains of the sky.

Thus, taking into account the above considerations, we may assume that *if a star of the Almagest has received no unambiguous identification, then its coordinates in the catalog are determined with a notable error.* Let us call such stars *doubtfully identifiable*, or *poorly identified*. So, *the number of poorly identified stars in a constellation provides an estimate for the number of poorly measured stars in this constellation.* This characteristic is undoubtedly of interest, because it shows which constellations were measured well and which were not. It is clear that the rate of doubtfully identifiable stars is a correct characteristic of the accuracy of measurements. In other words, we should find $(X/T) \cdot 100\%$ where T is the total number of stars and X is the number of doubtfully identifiable stars in the constellation.

The result accumulates a lot of preliminary work carried out by researchers of the *Almagest*. Since there were many researchers, we have grounds to assume that the average of their results gives a more or less reliable picture, free of biases of this or that specialist.

We have carried out the calculations and summarized the results in Table 2.2. The first column of the table indicates the domain of the sky that contains the constellation; recall that we have distinguished seven domains, *A, Zod A, B, Zod B, M, C* and *D*.

The table contains all 48 constellations mentioned in the *Almagest*; it also contains data on *informatas*, the list stars which were not attributed by Ptolemy to the constellations but indicated as lying near this or that constellation.

3. Seven domains in the star atlas of the *Almagest* and accuracy of measurements

We draw the following conclusions from Table 2.2.

Conclusion 1. *The seven domains as in Section 2 consist of the following constellations:*

Domain A: constellations 1–8 and 24–29.
Domain B: constellations 16–23 and 30–33.
Domain Zod A (contained in A): constellations 24–29.
Domain Zod B (contained in B): constellations 22, 23, 30–33.
Domain D: constellations 34–38, 47 and 48.
Domain C: constellations 39–46.
Domain M: constellations 9–15.

Conclusion 2. *Usually, informatas of the constellations (if any) are very poorly measured. In fact, only the following informatas are measured well: Ursa Minor (one star), Bootes (one star), Hercules (one star), Cygnus (two stars), Ophiuchus (five stars), Aquila (six stars), Aquarius (three stars), and Pisces (four stars), that is, only 9 of 22 informatas. The rest of the 13 informatas are measured very poorly. Indeed, we have 38% of poorly measured stars in the informata of Ursa Major, 50% in the one of Cepheus, 33.3% for Perseus, 36.4% for Taurus, 57% for Gemini, 75% for Cancer, 37.5% for Leo, 16.6% for Virgo, 44.4% for Libra, 66.7% for Scorpius, and 100% for Canis Major, Hydra and Piscis Austrinus.*

Table 2.2.

Constellation	Domain	Presence of an informata	Percentage of poorly identified stars			Number of stars in the constellation	
			"pure"	informata	total	"pure"	total
1. Ursa Minor	A	+	0	0	0	7	1
2. Ursa Major	A	+	3.7	11.4	38.0	27	8
3. Draco	A		0			31	0
4. Cepheus	A	+	0	7.7	50.0	11	2
5. Bootes	A	+	27.3	26.0	0	22	1
6. Corona Borealis	A		0			8	0
7. Hercules	A	+	10.3	10.0	0	29	1
8. Lyra	A		10.0			10	0
9. Cygnus	M	+	0	0	0	17	2
10. Cassiopeia	M		23.0			13	0
11. Perseus	M	+	3.8	6.9	33.3	26	3
12. Auriga	M		21.4			14	0
13. Ophiuchus	M	+	25.0	20.7	0	24	5
14. Serpens	M		0			18	0
15. Sagitta	M		0			5	0
16. Aquila	B	+	22.3	13.3	0	9	6
17. Delphinus	B		20.0			10	0
18. Equuleus	B		100.0			4	0
19. Pegasus	B		10.0			20	0
20. Andromeda	B		13.0			23	0
21. Triangulum	B		0			4	0
22. Aries	Zod B	+	0	0	0	13	5
23. Taurus	Zod B	+	21.2	25.0	36.4	33	11
24. Gemini	Zod B	+	5.6	20.0	57.0	18	7
25. Cancer	Zod A	+	0	23.0	75.0	9	4
26. Leo	Zod A	+	11.1	17.1	37.5	27	8
27. Virgo	Zod A	+	15.4	15.6	16.6	26	6
28. Libra	Zod A	+	0	23.5	44.4	8	9
29. Scorpius	Zod A	+	4.8	12.5	66.7	21	3
30. Sagittarius	Zod B		12.9			31	0
31. Capricornus	Zod B		3.6			28	0
32. Aquarius	Zod B	+	26.1	24.4	0	42	3
33. Pisces	Zod B	+	5.8	5.2	0	34	4
34. Cetus	D		22.7			22	0
35. Orion	D		8.9			38	0
36. Eridanus	D		26.4			34	0
37. Lepus	D		0			12	0
38. Canis Major	D	+	5.6	41.3	100.0	18	11
39. Canis Minor	C		0			2	0
40. Argo Navis	C		68.9			45	0
41. Hydra	C	+	16.0	22.2	100.0	25	2
42. Crater	C		57.1			7	0
43. Corvus	C		0			7	0
44. Centaurus	C		81.0			37	0
45. Lupus	C		100.0			19	0
46. Ara	C		100.0			7	0
47. Corona Australis	D		100.0			13	0
48. Pisces Austrinus	D	+	8.3	38.9	100.0	12	6

Thus, on the whole, the informatas of the Almagest concentrate poorly measured stars. The hypothesis seems appropriate (though not influencing our further considerations) that since informatas gather stars outside the main bodies of the constellations, no particular significance was attached to their measurement, especially when dealing with comparatively faint stars. Of course, the coordinates of bright stars, even though in an informata, were measured much more accurately (for example, the famous Arcturus is contained in the well-measured informata of Aquarius). However, as can be seen from Table 2.2, *the situation when the stars in an informat are measured much less accurately than the ones in the "main body" of a constellation is typical.*

Therefore it seems natural to separate the stars in the informatas from the stars of the constellations themselves (which, in fact, is done in the *Almagest*, where stars of informatas are gathered in a separate group under the title *Informata*), and consider the "main stars", that is, the stars contained in the main bodies of the constellations.

To make the picture more visual, we display the data of Table 2.2 in Figure 2.3; we place here two numbers inside every constellation, the first of which (numerator) is the percentage of poorly measured stars of the main body of the constellation, and the second (denominator) is the percentage of poorly measured stars in the constellation together with its informata. For constellations that have no informata, we omit the second number (leaving the fraction sign). The dotted line in Figure 2.3 delineates the Milky Way.

A close examination of Figure 2.3 reveals some interesting regularities.

To make the numerical data still more visual, let us look at Figure 2.4, in which the domains with the rates 0% to 5% are left white, the ones with the rate 6% to 10% are dotted, with the rates 11% to 20% are marked by oblique hatches, the domains with the rates 21% to 30% are double-hatched, and the domains within the rates 31% to 100% are black; so, the darker a domain in Figure 2.4, the worse it is measured. It is immediately obvious that many southern constellations in the domain *C* (to the right from the Milky Way) are very poorly measured. To the contrary, the constellations in the domain *A* are measured much better. The domain *B* is measured worse than the domain *A*. Some domains in Figure 2.4 are marked by the question sign; these are the domains of the modern starry sky which are not covered by the constellations mentioned in the *Almagest*. We may assume here that since the boundaries of constellations are not clearly defined in the *Almagest*, we can "expand" the constellations so that they cover the empty domains in Figure 2.4. We will not detail the procedure, because such domains are few, and they do not effect our conclusions.

Now let us calculate mean percentages of poorly measured stars over each of the seven domains. To this end, sum up the previously found rates for all constellations of a domain and divide the sum by the number of constellations in the domain. The results are displayed in Table 2.3.

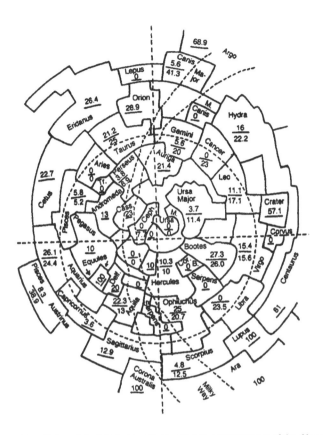

Figure 2.3. Rates of doubtfully identifiable stars in constellations of the *Almagest*.

Conclusion 3. *The domain A is measured better than the domains B, C, D and M (namely, 6.3% poorly identifiable stars in the main bodies of the constellations, and 12.6% in the constellations together with the informatas).*

Conclusion 4. *The domain B is measured worse than the domain A (19.6% poorly identifiable stars for the "pure" constellations and 19% for constellations with informatas).*

Conclusion 5. *The domain M (the Milky Way) is between domains A and B (10.5% for "pure" constellations and 10.3% for constellations with informatas).*

Conclusion 6. *The domains C and D are the worst measured (27.4% and 36.9% for D, and 52.9% and 53.6% for C).*

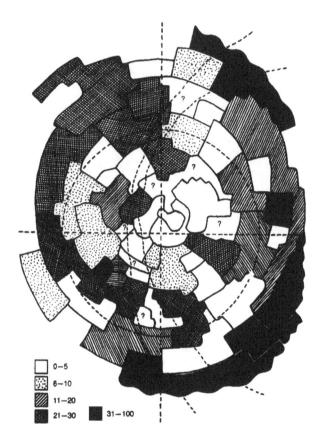

Figure 2.4. Rates of doubtfully identifiable stars in constellations of the *Almagest* (graphical representation).

Conclusion 7. *The best-measured domain is Zod A, the part of the zodiac to the right of the Milky Way (Gemini, Cancer, Leo, Virgo, Libra, Scorpius); the rate for "pure" constellations here is 6.2%.*

Conclusion 8. *The domain Zod B is measured with much lower accuracy than Zod A; here the rate is 11.6% for "pure" constellations. The domain consists of Sagittarius, Capricornus, Aquarius, Pisces, Aries and Taurus.*

To represent the information in Table 2.3 visually, we display it in Figure 2.5, where the white areas correspond to 0% to 10%, dotted areas to 10% to 20%, obliquely hatched to 20% to 30%, and doubly hatched to 30% to 100%.

Another display of this information is represented in Figure 2.6; the numbers of the 48 constellations are plotted along the horizontal line, in groups

Table 2.3.

Domains	Number of constellations	Number of constellations in the *Almagest*	Percentage of poorly identified stars in "pure" constellations	Percentage of poorly identified stars in constellations with informatas	Percentage of reliably identified stars in "pure" constellations
A	14	1–8, 24–29	6.3	12.6	93.7
B	12	16–23, 30–33	19.6	19.0	80.4
A – Zod A	8	1–8	6.4	8.1	93.6
B – Zod B	6	16–21	27.6	26.5	72.4
Zod A	6	24–29	6.2	18.6	93.8
Zod B	6	22, 23, 30–33	11.6	11.9	88.4
D	7	34–38, 47, 48	27.4	36.9	72.6
C	8	39–46	52.9	53.6	47.1
M	7	9–15	10.5	10.3	89.5

A, B, Zod A, Zod B, A – Zod A, B – Zod B, M, C, D. On the vertical the rate of doubtfully identifiable stars in "pure" constellations is plotted. We assign to each group of constellations a horizontal segment in Figure 2.6, at the level corresponding to the mean rate for this group. It is obvious from Figure 2.6 that the group *A*, consisting of *Zod A* and *A – Zod A* is measured best. The group *B* is disposed considerably higher in Figure 2.6, which corresponds to a worse accuracy of measurement.

The same information from the last row of Table 2.3 is also shown in Figure 2.7; here we plot along the vertical the rate of reliably identified stars in "pure" constellations. Clearly, this graph can be obtained by subtracting the values in Figure 2.6 from 100%.

Conclusion 9.   *The seven domains of the sky we have distinguished above differ in the accuracy of measurement of stellar positions.*

Conclusion 10.

a) *A further investigation of coordinates of stars in the Almagest should be based mainly on the stars of group A, as the best measured. This group has the minimal rate of poorly measured stars.*

b) *No conclusions should be based on a study of stars of groups C and D. The very high rate of poorly identified stars in these domains shows that these domains cannot be considered as well-measured. Probably, the distortion due to refraction of light beams was one of the reasons that hindered the measurements.*

c) *We get an opportunity to range the twelve named stars according to the reliability of measurement of their position. We should treat as the best measured the stars in the domain A and near it; there are nine such stars: Regulus, Spica, Vindemiatrix, Procyon, Arcturus, Aselli, Antares, Lyra (Vega). The stars Sirius*

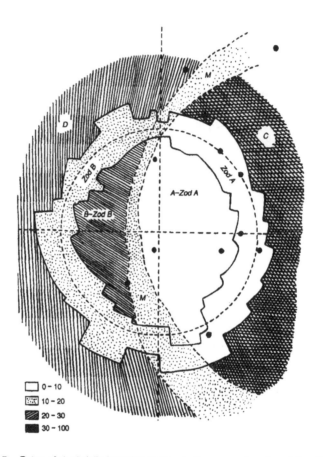

Figure 2.5. Rates of doubtfully identifiable stars in the seven domains of the *Almagest* (graphical representation).

(in the domain D), Aquila (in the domain B, at the left rim of the Milky Way), Canopus (outside the map) turn out to be "unreliable" as disposed in the "poorly measured" areas of the sky.

Remark. Apparently, Vindemiatrix should be excluded from the list of well-measured stars, because although it is identified well (in particular, does not appear in Table 6 of Ref. 22), the coordinates attributed to this star[22] are not based on an original manuscript of the *Almagest*. Peters writes about the coordinates of Vindemiatrix: "Greek authorities give 20°10′, the Arabs 15°10′ (the difference amounts to five degrees of arc! —*Authors*). Peters has adopted 16°0′ from Halma, who is copied by Baily, and he remarks that Halma

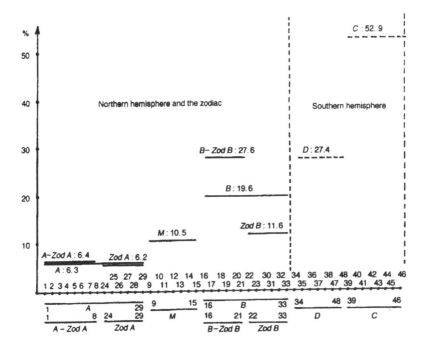

Figure 2.6. Rates of doubtfully identifiable stars in "pure" constellations of the *Almagest*.

gives no authority. It is clear that Halma took 16°0′ from Halley. It is of course correct, but is not supported by any manuscript" (Ref. 22, p. 104). It is clear that in this situation Vindemiatrix should be excluded from further treatment. Thus, only eight of twelve named stars are left among well-measured.

4. On possible distortions of stellar positions due to atmospheric refraction

Working with star catalogs, we should keep in mind the phenomenon of atmospheric refraction, which may distort essentially the positions of southern stars. The *refraction* is due to optical properties of the atmosphere; it affects the earthbound observations (as were all medieval observations). From the mathematical point of view, the atmosphere may be treated as a succession of concentric spherical strata; the density of the air is approximately constant within a layer, but varies from layer to layer. It is well known that light beams refract as they pass from a more dense stratum of air to a less dense (Figure 2.8); the more the difference of densities, the stronger is the refraction.

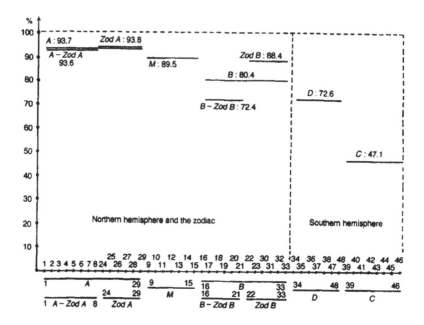

Figure 2.7. Rates of reliably identified stars in "pure" constellations of the *Almagest*.

As a result, the slope of the beam to the earth's surface increases, and the beam approaches the normal to the interface surface of the layers. Figure 2.9 depicts the earth's atmosphere, represented as the union of concentric layers, whose density decreases progressively with altitude. A light beam from a star A refracts as it passes the layers; as a result, it moves along a curved trajectory, the equation of which can be found (and this is done in the theory of atmospheric refraction). As shown in Figure 2.9, due to the refraction an observer on the surface sees the star in the beam OB, while the real position is in the beam OA'. Thus, *refraction increases altitudes of stars*. The closer the star to the horizon, the longer the beam travels in the atmosphere, hence the more is the apparent rise of the star. For higher stars the distortion is negligible. The following approximate expression is found in the theory of refraction for the refraction of zenith distances: the *zenith distance* ξ (the angle between the zenith at the vantage point and the direction towards the star) decreases by the magnitude approximately equal (for $\xi < 70°$) to

$$\rho \cdot 60'' \cdot \frac{B}{760} \cdot \frac{273}{273 + t°} \cdot \tan \xi$$

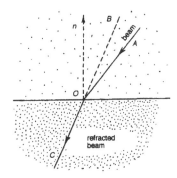

Figure 2.8. Refraction of a light beam at an interface between two media.

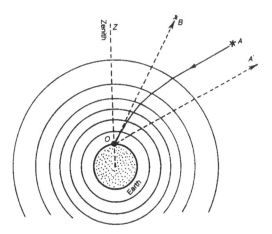

Figure 2.9. Refraction of a light beam in the atmosphere.

where B is the atmospheric pressure (in millimeters of mercury column, adjusted to $0°$ C), and $t°$ is the temperature of air (in Celsius degrees). It is obvious from this formula that $\tan \xi$ is the main factor influencing the refraction. For small zenith distances (i.e., for stars with large altitudes), $\tan \xi$ is small, so refraction is marginal. The closer a star to the horizon, the greater is $\tan \xi$, and consequently the more are the distortions of stars' positions caused by refraction. Probably, this is the reason why *in the Almagest (as well as in other star catalogs) southern stars (that is, stars with small altitudes) are measured much worse than stars of the Northern hemisphere.* We have already encountered this circumstance when we found out that *in the southern domains C and D of the star catalog of the Almagest the rate of doubtfully identifiable stars is considerably higher than that for domains A and B.* We should note that ancient astronomers were not aware of the existence of refraction, and even when refraction was discovered, an account of it was a highly nontrivial problem, coped with only as late as in the times of Tycho Brahe (probably, even later).

5. Analysis of distribution of informatas in the catalog of the Almagest

Table 2.2 contains data on distribution of informatas over constellations. It is clear from the table that only 22 of 48 constellations are supplemented with informatas. What does the presence (or the absence) of an informata for a constellation mean? Various points of view hereof are possible, but the following one appears to be most natural.

Conjecture. *The informatas were supplemented to the constellations considered to be most important. In other words, the presence of an informata is a sign of "especial attention" of the observer to the constellation.*

Apparently, some constellations were distinguished as especially important. We will not try here to find out the reasons for this distinction, because they are absolutely unessential for us (probably, they were of astrological nature, or some other). An especial attention was attracted by such constellations, so their stars were measured several times (which could bring about a higher accuracy). Moreover, having enlisted the stars that constitute the "figure" of the constellation (in our terminology, the stars of the "pure constellation"), the observer could add some "background" stars, which are not in the skeleton of the constellation, but are within the figure (or immediately near it). Thus the informatas could appear. As we already know, the stars of informatas, apparently treated as being of secondary importance, could be measured with lower accuracy than the stars of the "pure" (basic) constellation; nevertheless, the presence of an informata may be appraised as a sign of particular attention to the constellation.

Let us now look at the distribution of informatas over the starry sky of the *Almagest*.

In order to characterize the distribution quantitatively, we calculate for each constellation the percentage of stars which are included in the informata; thus, we calculate $c = (a/b) \cdot 100\%$ where a is the number of stars in the informata and b is the number of stars in the constellation together with the informata. So, for constellations without informatas, we have $c = 0$. Then we calculate the mean percentage for each of the groups A, B, M, etc. Thus, we find a numeric characteristic for each of the above domains of the sky,

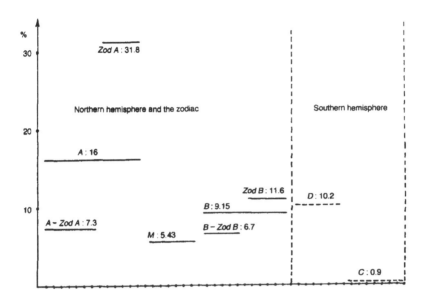

Figure 2.10. Rates of informatas (in percent) in various domains of the zodiac.

the mean rate of informatas in the domain. The results are displayed in Figure 2.10, following the pattern of Figure 2.6: we plot along the horizontal the numbers of constellations in the *Almagest*, grouped into the seven domains. Along the vertical we plot the mean percentage of stars in the informatas. As a result, a horizontal segment corresponds to each domain.

An important statement can be inferred from Figure 2.10:

Conclusion. *The distribution of "density of informatas" in the star atlas of the Almagest agrees with the distribution of percentage of doubtfully identified stars in pure constellations.*

This conclusion may be also formulated as follows: *The more attention paid by the observer to a constellation (that is, the more stars are included in the informata), the better the stars of the constellation are identified.*

Indeed, as is obvious from Figure 2.10, the density of informatas is the highest in *Zod A*, and the next is in *A*. More attention is paid to the domain *A* than to the domain *B*. In the Northern hemisphere, the least attention is paid to the domain *M*.

The least attention is paid to the domain *C* (Southern hemisphere). Although the domain *D* (also in the Southern hemisphere) acquired "comparatively much attention" (10.2%), the domain is measured worse. This is in no way surprising: the domains *C* and *D* constitute the *southern* part of the sky, which, as we have stressed several times, is inevitably measured with lower accuracy than the Northern hemisphere and the zodiac.

Thus, we draw the following statement from Figures 2.6 and 2.10:

Conclusion.   *The partition of the star atlas of the Almagest into seven domains is confirmed. The quality of measurements in each of the domains (we mean first of all the Northern hemisphere and the zodiac) is proportional to the "attention paid to the domain". The higher the density of informatas, the better the stars are measured (the higher is the rate of reliably identified stars). The lower the density of informatas, the lower is the rate of well-identified stars.*

6. On reliability of measurement of latitudes and longitudes in the *Almagest*

We begin this section with R. Newton's statement concerning accuracy of measurements in the *Almagest*. We think, however, that this statement is applicable within a much wider context; in a sense, this statement characterizes the situation around interpretations and readings of various historical documents. R. Newton tells of a well-known principle, "that we may call the immortality of error".

"We may state this principle in the following manner: Suppose that an error made by a writer A has somehow been published, and suppose further that a later writer B quotes and cites the error, accepting it as correct. The error then becomes immortal and cannot be eradicated from the scholarly literature. I do not maintain seriously that the principle has no exceptions. However there are distressingly many examples for which the principle is valid, and any reader can probably furnish his own examples" (Ref. 1, p. 161). Something like that is going on around interpretation and dating of the *Almagest*. An analysis of the traditional version attributing the *Almagest* to the beginning of the Christian era requires a thorough reanalysis of its contents, which constitutes a fairly complicated scientific problem. In this book we carry out a part of this work, and the reader can estimate the complexity of the problem. The main difficulty consists in the necessity to delve into the foundations of each scientific assertion or opinion, the overwhelming majority of which has been stated under the a priori (or silent) assumption that the *Almagest* was compiled in the beginning of our era. These "excavations" require an analysis of the original material, which in itself is quite arduous.

Let us now return to the question on the accuracy of measurement of latitudes and longitudes. As we have demonstrated in Chapter 1, the nature of ecliptic and equatorial coordinates implies a better measurement of latitudes in comparison with longitudes. Moreover, if we use, for example, an armilla, the errors may originate in an erroneous determination of the inclination of the ecliptic. The fact is that having determined the angle the ecliptic makes with the equator, the observer fixes it and uses the device for measuring coordinates, say, of stars, setting it with the help of this angle. Generally, an armilla may be set with the help of any object the latitude and longitude of which are known (for example, Ptolemy often used the moon); having done this, we can find coordinates of any heavenly object. But in this case the errors in determination of coordinates of the object used for setting automatically generate errors in the coordinates of the measured object (see Ref. 1, p. 145). Furthermore, we should keep in mind that in the case of the *Almagest*, we deal with the lists in which letters were used for digits, which might bring about confusions (and it actually did). For example, as noted by R. Newton (Ref. 1, p. 215), and Peters and Knobel[22], in ancient Greek notation, digits 1 and 4 were easy to confuse, because the digit 1 was denoted by a symbol that was an early version of α (in the time of Hipparchus), and the digit 4 was denoted by Δ. Clearly, the symbols Λ and Δ were easy to confuse. In connection with that we should make an important remark: *Our investigation is based on the version of the star catalog of the Almagest adduced in the work of Peters and Knobel*[22]. As R. Newton notes, "By a careful comparison of various manuscripts, it is often possible to detect errors that have occurred in the course of successive copyings and to correct them. Peters and Knobel ... have made an extensive study of the star catalog in the *Syntaxis* (= the *Almagest— Authors*), and theirs is probably the most accurate version of it that exists" (Ref. 1, p. 216). However, as Peters and Knobel corrected some digits in the *Almagest*, they of course used the a priori assumption that the catalog had been compiled in the beginning of our era (they obviously did not suspect other possibilities); consequently, from several possibilities they probably chose the ones that matched this assumption best. Thus their work on "correction" of numeric data in the catalog could lead to a "shift" of the data towards the beginning of our era. Realizing this possibility (of the "unintentional shift of the date"), we nevertheless used their version. The fact is that if the computational investigation of the catalog reveals the necessity of an essential redating, this must mean that this necessity is stable enough to stand the biased correction. This is a vivid example of the problems that arise. Strictly speaking, we should use the original manuscripts of the *Almagest*; possibly, they contain numeric data that were ignored by later researchers as contradictory to the a priori dating assumption.

Thus, we come to the necessity of a complete revision of the star catalog of the *Almagest* and of addressing ourselves to the original (by the way, not easily accessible) materials of ancient manuscripts in order to restore the real data of the catalog, free of distortions introduced by the researchers biased for the traditional dating. This, probably, is an example of "immortality of error". Anyway, we will stop our discussion of the subject here, and in the sequel we use the version of Peters and Knobel.

Even if the biased correction of data was done, we may suggest that it could affect only a minority of the 1025 stars in the catalog. Thus, our approach (we analyze families of stars rather than individual stars) must compensate for this correction. Therefore we did not expect major errors originating in the use of the canonical version of Peters and Knobel (which also has the advantage of being accepted as fundamental by other researchers).

In our estimate of reliability of altitudes and longitudes in the *Almagest*, we used the detailed analysis of the subject carried out by R. Newton in the extensive Chapter IX of his book[1]). Of course, we omit details and confine ourselves to quoting his results.

a) *Latitudes*. "The latitudes in the star catalogue were obtained by measurement, almost surely made by a single observer using a single instrument ... " (Ref. 1, p. 255). Furthermore, "The latitudes in the star catalog have not been altered from the observed values except perhaps by scribal accident" (Ref. 1, p. 252). In R. Newton's opinion, the latitudes in the catalog provide a reliable material, obtained by Ptolemy himself or some of his predecessors (presumably, by Hipparchus). This agrees perfectly with our above considerations showing that measuring latitudes is a simpler procedure than measuring longitudes and hence latitudes are more reliable data.

b) *Longitudes*. A different picture emerges here. " ... The longitudes were not obtained by any plausible observing process. ... The longitude values were fabricated by the process that has been described" (Ref. 1, p. 252). And further, "The longitudes in the star catalog cannot possibly be the result of observations" (Ref. 1, p. 252). First of all, as we have noted, measuring ecliptic longitudes is a far more complicated and delicate procedure than that of latitudes; besides, it is deemed that the longitudes of stars in the *Almagest* are reduced to 137 AD. A reduction of this kind (to a given date) is realized by merely adding a common constant to the ecliptic longitudes of all stars; the constant is proportional to the magnitude of precession and depends on the particular shift to the past the compiler wanted to apply to the data. R. Newton conjectures that the original longitudes (obtained experimentally by an unknown observer) were later recalculated by somebody else for an unknown reason. Here is the fundamental conclusion of R. Newton, which he draw from an analysis of occurrences of fractions of degrees of arc in the catalog: "The longitudes have been altered by adding an integral number of degrees, plus 40' " (Ref. 1, p. 252). This operation, addition of an integral number of degrees and a fraction, makes feasible an arbitrary change of the age of the catalog (recall that a similar manipulation with the latitudes is impossible, or at least much more difficult). To determine the particular magnitude of the shift from an analysis of the *longitudes alone* is impossible; this is also noted by R. Newton: "The distribution of the fractions by itself cannot tell us the integer part of the amount that Ptolemy added to the original longitudes" (Ref. 1, p. 253). Besides this trivial operation of shifting all longitudes by an unknown number of grades, R. Newton finds the traces of a more delicate recalculation of the longitudes (Ref. 1, pp. 246–250). Thus, somebody had done a significant change of the originally observed longitudes of stars, so the set of longitudes now at our disposal is not the result of observations, but of a processing (possibly, fairly complicated) of observational data. N. A. Morozov conjectured that this processing had been carried out for artificially "making the catalog more ancient". We will not try to find out the purposes of this recalculation, and will confine our further study to the altitudes.

In conclusion, another résumé of R. Newton: "The longitudes tell a quite different story from the latitudes. The distribution of the longitude fractions does not come from any possible body of observations, whether they were made by using one instrument or more, and whether they were made by one observer or more" (Ref. 1, p. 248).

In traditional history of astronomy, the following simple method is often used for dating catalogs. The ecliptic longitudes in the catalog are compared with the modern ones, and the difference between them (approximately the same for all stars) is divided by the magnitude of precession (approximately 50' per century); the resulting ratio gives the difference between the dates of compilation of the new catalog and the ancient one. In particular, with the help of this method, for example, the ecliptic latitudes in the Greek edition of the *Almagest* of 1538 have been attributed to the beginning of our era. It is *silently presumed* in this method that the compiler of the ancient catalog counted the ecliptic longitudes from the spring equinoctial point (of his time). If that is really so, then the variation of longitudes accumulated by our time may be interpreted as the effect of precession, and the above algorithm provides the approximate date of compilation of the catalog. It is important, however, that *not all ancient authors took the spring equinoctial point as the zero point for longitudes.* It should not be presumed that ancient astronomers counted longitudes exactly as modern astronomers do. Let us consider, for example, the famous *Historia Universalis Omnium Cometarum* by S. de Lubienietski, 1681. This book is known to be medieval; it contains many comets up to 1860, and the author belongs to the medieval astronomic school, where, as it seems, the system of base points and the rules for compilation catalogs had been unified. However, in his star maps, Lubienietski counts longitudes from the meridian through γ Arii; as a result, all longitudes he indicates are approximately 7° less than that in the Greek edition of the *Almagest* of 1538 (see, for example, comparative tables and maps in Ref. 4, vol. 4, pp. 233–234). If we now assume (as usually done dating the *Almagest*), that Lubienietski counts longitudes from the equinoctial point, we will have to attribute him and his book to the 5th century BC! This ludicrous result shows that we should be very cautious about the above dating procedure. We conclude also that *even as late as in the 17th century no conventional zero point for longitudes was established.* Lubienietski counted longitudes from the star in Aries, the author of the Greek edition of the *Almagest* from a point 6°40' off γ Arii, etc. Apparently, each medieval author had his own point of view on the appropriate zero point for longitudes (clearly, the choice of the point is purely conventional). A more impressive example is the catalog of Copernicus, who also counted longitudes from γ Arii (in fact, Lubienietski followed the tradition of Copernicus): γ Arii is the only star in the catalog that has longitude 0^{25}. Consequently, if we apply the above method to Copernicus, we will have to attribute him far into antiquity. Thus, the difference of ecliptic longitudes cannot serve as a base for dating catalogs. The inconsistency in the choice of zero points for longitudes is in fact quite natural. At the dawn of astronomy many often rival schools existed, which used various approaches to compilation of catalogs. Possibly, the schools used various traditions, including specific choice of basic point (zero points) for various astronomic characteristics. Some of them counted longitudes from the equinoctial point, the other from γ Arii, the third .... The choice could be determined by astronomic, religious, etc., considerations. Only when astronomy became an international science was the unification of astronomic language reached, in particular, the choice of the equinoctial point was fixed. By the way, from the observational point of view this point is fictitious, moreover, it moves along the sky, so it is impossible to fix it by indication of a nearby bright star. So it is not surprising that some medieval astronomers chose to use for reference an *observable* star (say, γ Arii).

Since we study the star catalog of the *Almagest* (and other medieval star catalogs), formally we "do not know" the point the longitudes are counted from therein (the catalog itself contains no indication on this point). Of course, it is traditionally accepted that the text of the *Almagest* (not the catalog) contains some indications on the choice of the equinoctial point; however, using this would mean that we involve some external information (not contained in the catalog). As for dating the written text of the *Almagest*, this is a separate question that has acquired no final solution (see Refs. 1,4). Therefore we will not rely upon the ambiguous data, requiring additional examination. It is relevant to note the existence of several star catalogs (among them, the catalog of the *Almagest* and Al Sûfi's catalog) that only differ in longitudes. More precisely, the latitudes given in the catalogs coincide, while all longitudes differ by a constant (in the two catalogs, by $12°42'$). An obvious conclusion is that at least one of the catalogs is not compiled from observations, so it would be clearly irrelevant to use longitudes it contains for dating.

7. Peters' sine curve and the latitudes in the *Almagest*

After remarks in the previous section, it makes sense to concentrate mainly on the study of latitudes of stars in the *Almagest*. Here we encounter at once an interesting phenomenon, unexplainable within the frames of traditional studies of the *Almagest*, which we call *Peters' sine curve*. The phenomenon is described by Peters in Ref. 22, where he carried out an analysis of the distribution of mean error in the latitudes of stars of the *Almagest* considered

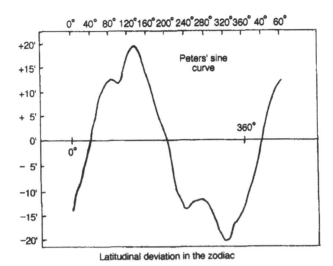

Figure 2.11. The sine-like dependence of latitudinal errors of longitudes, discovered in the *Almagest* by Peters; so far, this dependence has received no explanation.

as function of the longitudes. He computed the positions of zodiacal stars for 100 AD, that is, for the purported date of compilation of the catalog. Then, for each zodiacal star A_i he calculated the latitudinal deviation $\Delta_i = B_i - b_i$ where B_i is the latitude as given in the *Almagest*, and b is the computed value of the latitude in 100 AD. Thus, Δ is "Ptolemy's error" in the latitude of the ith star (under the assumption that the catalog was compiled about 100 AD). Then Peters divided the ecliptic into $10°$-long sectors and found the mean value of the deviation for each sector (over zodiacal stars of the *Almagest* in this sector). The ensuing values differed for different sectors; as a result, we obtain a graph that exhibits the behavior of the deviation as a function of latitudes; see Figure 2.11. The curve resembles the sine curve with the amplitude about 20 minutes of arc. We can select the sine curve that approximates the curve in Figure 2.11 best (within the class of all sine curves); the resulting sine curve is the *Peters' sine curve*.

The fact of the existence of this curve is difficult to explain on the basis of traditional views of the *Almagest*; at least, we have never seen a published attempt of reasonable explanation of this obviously periodic effect.

We should note that Ref. 22 exposes no details of Peters' calculations behind the curve; in particular, it is not clear what particular stars were used for the calculations. It is only known that Peters did not take into account *all* zodiacal stars of the *Almagest*. Therefore we have carried out the computation of the curve; see the discussion of our results, conclusions and comments in the

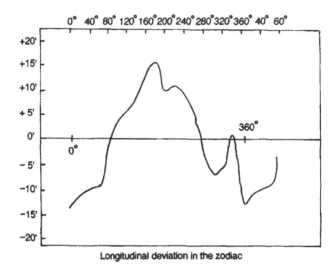

Figure 2.12. The dependence of longitudinal errors of longitudes, discovered in the *Almagest* by Peters.

sequel. Here, we only note that our results provide a complete explanation of the strange curve.

Remark. Together with latitudes, Peters examined from a similar point of view the longitudes. In other words, he computed the mean longitudinal deviation (in 10°-long sectors) and obtained a graph similar to that for the latitudes, see Figure 2.12. The graph exhibits the behavior of the mean longitudinal deviation as a function of the ecliptic longitude. This graph does not resemble any sine curve; it also has lesser amplitude and two clearly defined local maximums. Possibly, the obviously irregular nature of this curve is a consequence of the unknown recalculations of ecliptic longitudes mentioned by R. Newton (see the previous section). Anyway, as we have already stated, the longitudes are hardly reliable data, so we see no reason for a more detailed study of the graph in Figure 2.12. An informative analysis of the graph, probably, will not be possible before we restore the mechanism of the recalculation of the longitudes, which must be very difficult.

Chapter 3

The Attempts to Date the *Almagest* with the Help of Simplest Procedures, and Why They Fail

1. An attempt to date the *Almagest*: Comparison with computed catalogs in fast stars

1. We have suggested in Chapter 1 an algorithm for computing stellar positions to the past. Having at our disposal the collection $\{K(t)\}$ of catalogs computed for various dates, we now can try to estimate the date of compilation of the catalog of the *Almagest* t_A from comparison with the computed catalogs; it is natural to treat as an estimate for t_A the value of t for which $K(t)$ agrees best with the catalog of the *Almagest*. Not fixing criteria of good agreement so far, let us look at how to compare the catalog of the *Almagest* with $K(t)$. In order to compare, we first need to reduce the two catalogs to the same coordinate system; to that end we bring the ecliptic of the *Almagest* into coincidence with the ecliptic of $K(t)$, that is, with the true position of the ecliptic in the year t. This enables us to compare latitudes; to compare longitudes, we need to fix the spring equinoctial point in the ecliptic of the *Almagest* (for the year t). We choose this point so that the mean longitudinal error over zodiacal stars of the *Almagest* will be equal to zero (we use the identifications of stars of the *Almagest* suggested in Ref. 22). To calculate this point is not difficult: it is known [22,26] that $t = 18.4$ (in 60 AD) was at the beginning of the zodiacal sign Aries, and that it moves at the rate of approximately 49″ a year (the velocity of precession).

This choice of the equinoctial point, though statistically best, introduces an error. We could avoid this error by confining ourselves to the latitudes;

we will do so in Chapters 3–5; the arguments in this section are preliminary.

2.   We will choose for comparison nine fastest stars among those indicated in the *Almagest* (according to Ref. 22); these are the stars whose velocities of proper motion exceed $1''$ per year. The stars are α Cent ($4''.08$ per year), o^2 Eri ($3''.68$ per year), α Boo = Arcturus ($2''.28$ per year), τ Cet ($1''.92$ per year), α CMa= Sirius ($1''.33$ per year), γ Ser ($1''.32$ per year), ι Per ($1''.27$ per year), α CMi = Procyon ($1''.25$ per year), η Cas ($1''.22$ per year). All the nine stars are present in the *Almagest*, and are traditionally identified [22] with the stars with the Baily numbers 969, 779, 110, 723, 818, 265, 196, 848, 180. We depict each of these stars except α Cent, which is very far to the south and whose coordinates are given in the *Almagest* with the huge error of 8°, as white circles in Figures 3.1–3.8; near each circle we write the corresponding Baily number of the star in the *Almagest*. Thus we get eight small domains in the star map of the *Almagest* (we used the coordinates of the *Almagest*), each of which contains one of the fast stars.

Let us now depict the true positions of the eight stars in the star map relevant to Ptolemy's catalog; the positions depend on t (as well as the identification of the catalog with $K(t)$), so we actually obtain eight trajectories that the stars pass with variation of t. These trajectories are shown in Figures 3.1–3.8.

Let us consider the moments $t_1, \ldots, t_8$ when the stars are nearest to their positions given in the *Almagest*. The eight values of t are different; if all of them, or at least most of them were close to each other and to some average value t^*, that would be a strong reason to suggest that the catalog was compiled about t^*. However, *this is not the case: the eight values are scattered chaotically in the interval* $-70 \le t \le 30$, *that is, from 1000 BC to 9000 AD!* The following table gathers the results of the calculation.

In fact, this spread of individual estimates of dates t_i is not surprising. Each of the eight stars is represented in the *Almagest* with an error. An estimate for this error is provided by the mean angular deviation of stellar positions given in the *Almagest* from the true positions. Since the overwhelming

Star	The moment of the closest approach to the star of the *Almagest*	minimal distance
Arcturus (α Boo)	900 AD	40′
Sirius (α CMa)	400 AD	10′
Procyon (α CMi)	1000 AD	20′
o^2 Eri	50 BC	5′
η Cas	1100 BC	40′
ι Per	9700 AD	70′
τ Cet	220 AD	15′
γ Ser	700 AD	80′

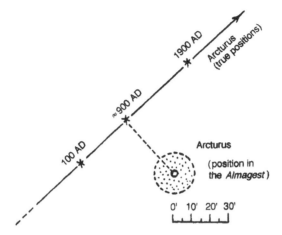

Figure 3.1. The trajectory of the true motion of Arcturus, in comparison with its position given in the *Almagest*. The dashed line depicts the 10′ neighborhood of the position given in the *Almagest*.

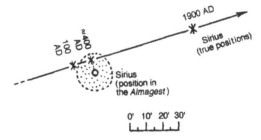

Figure 3.2. The trajectory of the true motion of Sirius, in comparison with its position given in the *Almagest*. The dashed line depicts the 10′ neighborhood of the position given in the *Almagest*.

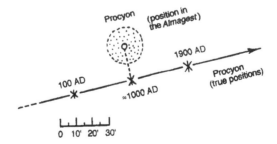

Figure 3.3. The trajectory of the true motion of Procyon, in comparison with its position given in the *Almagest*. The dashed line depicts the 10′ neighborhood of the position given in the *Almagest*.

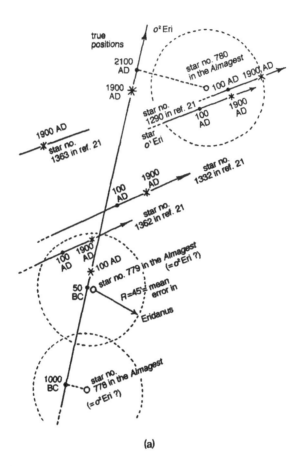

(a)

Figure 3.4 (a). True trajectory of the proper motion of o^2 Eri, in comparison with positions of stars of the *Almagest* with which it may be identified. The dashed lines depict 45' neighborhoods of the stars of the *Almagest*. The choice of the size is due to the error in stellar coordinates characteristic for description of Eridanus in the *Almagest*.

majority of stars are almost fixed, the mean error depends but little on the epoch for which the coordinates were computed. To find the mean error, we use the table in Ref. 22 that compares stellar positions given in the *Almagest* with the true positions in 130 AD. Around each point corresponding to a fast star, draw the circle of the radius equal to the mean error over the constellation that contains the star (see Figures 3.4–3.8). The projection of the circle on the computed trajectory of the star provides an estimate for possible deviation of the date t_i from the true date of compilation t_A. For the named stars

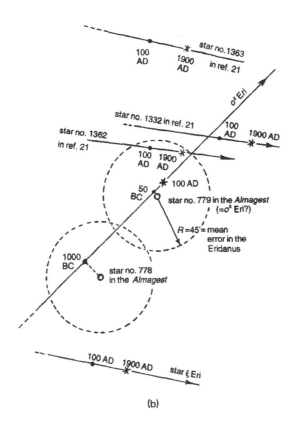

Figure 3.4 (b).   See the text at Fig. 3.4 (a).

(Arcturus, Sirius and Procyon), we chose 10', the value of the scale division
of the catalog of the *Almagest* as the radius of the circles (Figures 3.1–3.3).

3.   In this situation, the question naturally arises whether it is possible to
treat for some reason some of the eight stars as more reliable than the other.
If yes, then we should use these stars for dating, neglecting the rest. It is
natural to choose as reliable the stars that are measured with best accuracy.
But how can we distinguish such stars? Of course, we could do this using the
last column of the above table, that is, assuming, say, that the coordinates of
o^2 Eri were measured to the accuracy of 5', and the coordinates of Arcturus
to the accuracy of 40'. This approach was used in Ref. 27, an exposition of
an attempt to date the *Almagest*; apparently, the same nine fast stars were
used. The resulting date is close to traditional, about 50 BC (see the above
table). However, the following questions arise at once: How could it happen

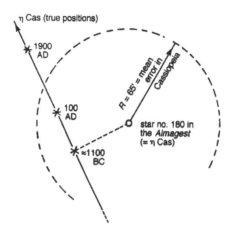

Figure 3.5. True trajectory of the proper motion of η Cas, in comparison with positions of stars of the *Almagest* with which it may be identified. The dashed lines depict 65′ neighborhoods of the stars of the *Almagest*. The choice of the size is due to the error in stellar coordinates characteristic for description of the star o^2 Eridani in the *Almagest*.

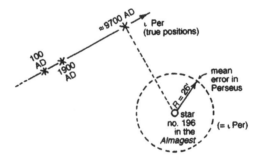

Figure 3.6. True trajectory of the proper motion of ι Per, in comparison with positions of stars of the *Almagest* with which it may be identified. The dashed lines depict 65′ neighborhoods of the stars of the *Almagest*. The choice of the size is due to the error in stellar coordinates characteristic for description of the star o^2 Eridani in the *Almagest*.

that *all three stars of the first magnitude,* all three having names in the catalog (Arcturus, Sirius and Procyon) had been measured *very roughly* (with errors about 1° if we assume 50 BC as the date of compilation — see Figures 3.1–3.3), while the faint and hardly visible o^2 Eri (whose magnitude, according to modern measurements is 4.5) had been measured with an exclusive accuracy (within 5′)? In fact, the stars like Arcturus, Sirius, Procyon, Regulus and Spica (that is, the stars that are named in the *Almagest*) apparently were used as

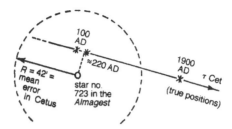

Figure 3.7. True trajectory of the proper motion of τ Cet, in comparison with positions of stars of the *Almagest* with which it may be identified. The dashed lines depict 65' neighborhoods of the stars of the *Almagest*. The choice of the size is due to the error in stellar coordinates characteristic for description of Cetus in the *Almagest*.

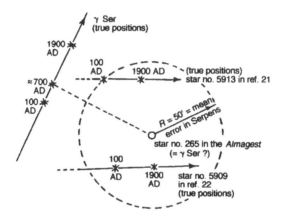

Figure 3.8. True trajectory of the proper motion of γ Ser, in comparison with positions of stars of the *Almagest* with which it may be identified. The dashed lines depict 65' neighborhoods of the stars of the *Almagest*. The choice of the size is due to the error in stellar coordinates characteristic for description of Serpens in the *Almagest*.

reference points for observations, and the accuracy of measurement of these stars had to be especially high (see, for example, Ref. 25). As for o^2 Eri, this is an ordinary star, surrounded by equally faint stars. In the *Almagest*, the star 779 (traditionally identified with o^2 Eri) is described merely as "the middle one of these". Furthermore, the other question arises after Figure 3.4: Why is the star 779 of the *Almagest* identified with o^2 Eri? Of course, this identification could be obtained from the fact that the coordinates of o^2 Eri and the star 779 of the *Almagest* agree best (say, better than the coordinates of o^2 Eri and the star 778). But then the identification o^2 Eri = 779 of the *Almagest* relies heavily on the date attributed to the *Almagest*! For example, should we "know" that the *Almagest* had been compiled in 1000 BC, we would identify o^2 Eri with the star 778 of the *Almagest*, and then date the *Almagest* from the minimal possible distance between the two stars, getting 1000 BC as the date. Note that in this case the coordinates in the *Almagest* and the coordinates of real stars would even agree much better, which is obvious from Figure 3.4. If we "knew" that the *Almagest* had been written in 1500 AD, then we would identify o^2 Eri with the star 780 of the *Almagest* and then, using the above approach, attribute the catalog to the late Middle Ages or even to the future — see Figure 3.4. Thus, we merely have a vicious circle.

However, even if we exclude o^2 Eri, no date for the *Almagest* can be derived from the remaining eight stars — the scatter of the dates provided by the stars is too wide. Even the dates from the three stars of the first magnitude, Arcturus, Sirius and Procyon, are scattered from 400 AD to 1000 AD (see the above table). Furthermore, we should keep in mind that the dates (for example, 900 AD for Arcturus) are only the dates of the closest approach of the true position of the star with the position given in the catalog; we should also take into account the intervals about these dates in which the approach lies within the accuracy of measurement. Still worse, we have to use individual errors of measurement rather than mean errors.

Let us now formulate our conclusions.

1) Using the coordinates of a star for dating the *Almagest*, we must first check that its identification with the star in the modern sky does not depend upon a particular date ascribed to the catalog; otherwise we get into a vicious circle.

2) Since the displacements even of the fastest stars accumulated within the historical period of time are rather small (see Figures 3.1–3.8), the stars of the *Almagest* that are measured sufficiently well should be selected for dating purposes. A star moving at the rate of 2″ per year shifts by mere 3.33 minutes of arc in a century. Therefore, if we want to get a date for the *Almagest* from a particular fast star with accuracy within 300 years, we should be sure that the star is represented in the *Almagest* with accuracy within 10′. It is known that the real accuracy of the catalog is much worse [22]. On the other hand, the stars represented with accuracy worse than 20′ are practically useless (the dating interval they provide is 1200 years long). The selection of stars that are measured by Ptolemy comparatively well is discussed in Chapters 5 and 6.

2. An attempt to date the *Almagest*: Comparison with computed catalogs in fast and named stars

1. In the previous section we have shown that the comparison of the *Almagest* with the computed catalogs $K(t)$ in eight fastest stars leads to no value t^* at which the *Almagest* and $K(t)$ agree best: each star gives its own

value t_i^* for t^*, and these individual values are scattered in an interval several thousand years long. It might happen, however, that if we extend the sampling and consider more stars, then we get a collection of individual dates $\{t_i^*\}$, a considerable part of which is grouped in a more or less narrow interval (an interval of about 500 years in length would already be good for us); then we would be able to get some information about the true date of Ptolemy's observations t_A. Furthermore, extending the sampling, we get the opportunity to apply standard methods of statistics to estimation of t_A.

For dating purposes, only sufficiently fast and sufficiently well measured stars may be useful. The two properties, fast motion and accuracy of measurement are, so to say, complementary: the faster is the star, the greater error of measurement is admissible.

The above considerations lead to the following choice: let us include in the sampling the stars that are sufficiently fast (such that at least one of the velocities in equatorial coordinates α_{1900} or δ_{1900} exceeds $0''.5$ per year), and "famous", or named stars, that is, the ones that have (by now) proper names. Of course, some of them received names after compilation of the *Almagest*, but, on the one hand, names of stars apparently were not lost with time (although could change), so the stars that were named in the time of compilation of the *Almagest* must remain named now; on the other hand, the fact that a star has finally acquired a name means that it had a special significance in ancient or medieval astronomy (the names were mainly attributed to stars in antiquity and the Middle Ages). It is natural to suppose therefore that Ptolemy gave special attention to such stars, and measured their coordinates more thoroughly.

In this section we take the interval $0 \le t \le 30$ (1100 BC to 1900 AD) as the a priori dating interval.

2.   Thus, let us consider the union of the lists of fast and of named stars (see Tables 1.1 and 1.2) and select those stars which are present in the *Almagest* (according to Ref. 22). The ensuing list contains 80 stars. For each of them, we compute the trajectory in the coordinates of the *Almagest*, as we did for the eight fastest stars in the previous section. Recall that we fixed t as the a priori date and found true positions of stars in the year t (in ecliptic coordinates of the year t) which we depicted as a point in the star map of Ptolemy (that is, in the map constructed according to the *Almagest*). Varying the value of t over the interval we have fixed, we make the point move along the atlas; consequently the distance varies between this point and the position of the star given in the *Almagest* our star is identified with (recall that we use so far the identifications of Ref. 22). Suppose the distance attains the minimum value at $t = t_i$; in the previous section we called the moment t_i the *individual date provided by the star*. A deviation of t from t_i to either side leads to an increase of the distance between the computed position of the star and the position given in the *Almagest*. We assign to each star in our list the *individual*

dating interval $[t_{i1}, t_{i2}]$, the set of all t for which the distance does not exceed 30′. Generally, this interval can be empty (if even the minimal distance at $t = t_i$ is greater than 30′). The value $t_i = t$ is the center of this interval — see Figure 3.9. The value 30′ for the maximal admissible distance between the star of the *Almagest* and the true (= computed) position is chosen so that the distances do not exceed it for a majority of stars of the *Almagest*. Indeed, if we assume that the mean square deviation in angular distance for the *Almagest* is about 40′ (which agrees with estimates of Refs. 22 and 28), then more than half of the stars must be represented to an accuracy within 30′ (if we assume normality of distribution and independence of errors in positions of individual stars; our argument is quite rough, so deviations from these assumption will not affect our conclusions).

The resulting set of intervals is shown in Figure 3.10. Time scale from $t = 0$ (1900 AD) to $t = 30$ (1100 BC) is plotted along the vertical. In each interval, the center is marked (corresponding to the optimal individual date t_i) and the points that correspond to distances 10′ and 20′ (see Figure 3.9); the parts of intervals corresponding to less distances than 10′ are marked by bold lines; the ends of intervals are marked by arrows. No interval is assigned to many stars from our list; this means that either the interval is empty (the distance between the true position of the star and the one given in the *Almagest* always exceeds 30′), or that the interval lies outside the a priori interval $0 \leq t \leq 30$, or contains the a priori interval. In the latter case, the coordinates were apparently measured well (to an accuracy within 30′), but they provide no refinement for the date within the a priori interval.

The Baily numbers of the stars in the *Almagest* whose 30′ approach intervals cover the a priori dating interval $0 \leq t \leq 30$ are 35, 36, 163, 197, 222, 316, 318, 376, 768. For some stars the individual dating intervals extend outside the a priori dating interval, so we only expose their common part in Figure 3.10. Near each interval, we write the Baily number of the star, and after the equality sign, modern denotation of the star identified in Ref. 22 with the star in question. The dashed line marks the level $t = 18$, the traditional date for the *Almagest* about 100 AD.

3. It is obvious from Figure 3.10 that *no values of t that belong to all individual dating intervals exist.* If we raise the value of admissible accuracy (which we took as 30′), thus expanding the individual dating intervals, the point where they first intersect is about $t = 12$ (700 AD), at an accuracy of about $60′ = 1°$. If we increase the admissible accuracy still more, the interval which is the intersection of all individual dating intervals expands in both direction from the value $t = 12$.

However, we cannot accept the value $t = 12$ as a date for compilation of the *Almagest*. Indeed, the fact that the intersection first appears at the value of accuracy of about 1° means that our collection of stars contains very poorly measured stars. We do not know the exact accuracy of measurement of

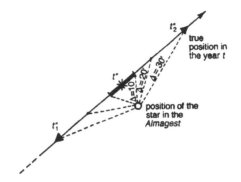

Figure 3.9. Determination of the individual date for the catalog from a fast star.

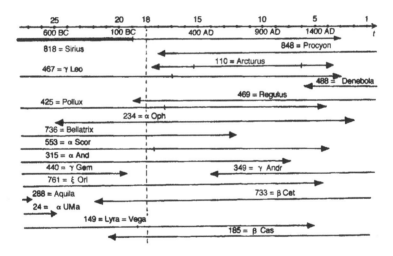

Figure 3.10. The collection of nontrivial individual dating intervals for all fast and named stars. The points are marked in each interval where the distance between the star of the *Almagest* and the computed position is 10′ and 20′.

particular stars in the *Almagest*; if we estimate the accuracy from below with the help of sampling mean square error at $t = 12$, then we will have to take a very large value, about 2°, for the admissible accuracy. At this accuracy, the intersection of all individual dating intervals covers the time from 500 BC to our day. Furthermore, the value $t = 12$ is *unstable*, in the sense that minor alterations of our collection of stars (obviously, chosen fairly arbitrarily) bring about considerable variations of the date.

4.   Let us now apply a similar procedure to the collection of fast and named stars, only using as a distance between a stellar position in the *Almagest* and the computed position, the *latitudinal deviation*, the length of projection of the segment connecting the two points on the meridian of the coordinate system of the *Almagest* (Figure 3.11). As we know, longitudes are given in the *Almagest* with much lower accuracy than latitudes, (see, for example Ref. 22 and Chapter 2), so we do not use them.

Figure 3.12 shows the collection of intervals for the case when latitudinal deviation is taken for distance. Again, we do not show intervals that cover the interval $0 \leq t \leq 30$ (1100 BC–1900 AD); the numbers in the *Almagest* of such stars are 1, 35, 36, 78, 111, 149, 163, 189, 222, 234, 287, 288, 315, 316, 318, 349, 375, 393, 410, 411, 424, 467, 469, 510, 713, 733, 760, 761, 768, 812, 816. A comparison of Figures 3.12 and 3.10 shows that the latitudes of stars are given in the *Almagest* with a much better accuracy than their positions (determined both by latitudes and longitudes); in particular, Figure 3.12 contains many more intervals than Figure 3.10 (where the corresponding intervals turned out to be empty). The individual dating intervals of all stars but two (935 = $2g$ Cent and 940 = $5q$ Cent) first intersect at $t = 12$ (700 AD) at the value of accuracy 40′ (in latitude).

Generally, although switching from angular deviations to latitudinal deviations improves accuracy, and hence allows more accurate statistical conclusions, the ensuing dating intervals are still too large: they cover the interval $4 \leq t \leq 20$ (1000 BC–1500 AD). These intervals provide no nontrivial information about the date of Ptolemy's observations.

3. An attempt to date the *Almagest* from comparison of stellar configurations

1.   In the previous sections, we tried to date the *Almagest* considering various stellar configurations that vary with time because of proper motions of the stars they consist of. For each individual star of the configuration, we compared its position in relation to the sphere of fixed stars with the position given in the *Almagest*; this comparison required use of Newcomb's theory, providing a description of motion of the ecliptic frame of reference in the sphere of fixed stars.

Let us now look at what we can obtain from a method that does not use Newcomb's theory. The idea of the method is quite simple: we compare not the positions of stars in the "real" (computed) sky with the positions in the *Almagest*, but the geometry of varying real stellar configurations with the geometry of stellar configurations in the *Almagest*. This comparison only requires knowledge of proper motions of stars, but not of Newcomb's theory. The proper motions of stars are now measured with very high accuracy on the basis of telescopic observations.[22,29] The only information we need in this

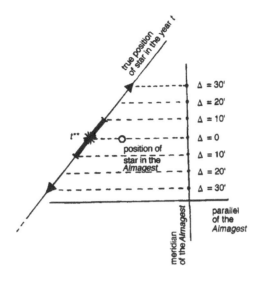

Figure 3.11. Determination of the individual date for the catalog from a fast star; only latitudinal deviations considered.

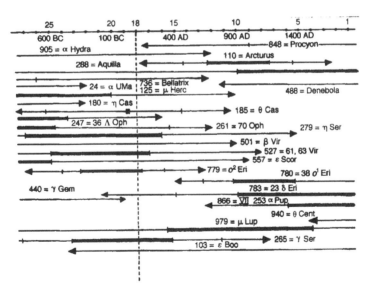

Figure 3.12. The collection of nontrivial individual dating intervals for all fast and named stars (only latitudinal deviations considered). The intervals in bold correspond to latitudinal deviations below 10′.

section is the velocities of proper motion of stars and the table of identification of stars of the *Almagest* with the stars of modern catalogs. We use the
identifications of Ref. 22, dropping the cases indicated as doubtful.

Although Newcomb's theory is very precise (its errors are several orders
of magnitude less than the scale division of the *Almagest*), from the computational point of view, configurations are more convenient, first of all for the
simplicity of calculations. Though the abandonment of Newcomb's theory has
some drawbacks, confining ourselves to comparison of configurations, we lose
the possibility to separate coordinates (and study separately, say, latitudinal
deviations), and are left to compare angular distances alone.

2.   We will still compare the positions of fast stars in the real sky with
the positions given in the *Almagest*, but now we will compare positions in
relation to a certain set of reference stars, distinguished both in the real sky
and in the *Almagest*. We include in the set the stars which either have proper
names (Aldebaran, Shiat etc.) or stand out for their brightness among the
nearby stars. We did not include the stars the positions of which could be
distorted by refraction. The total number of the reference stars is 42; see the
Appendix for the list (Table Ap. 1). Some of these stars have notable velocity of
proper motion (Arcturus, Sirius, Procyon, Capella, Aquila, Denebola, Caph
and Regulus). Thus, the true position of a star is determined in reference to a
basis, which also moves, and the ensuing "dynamic" picture is compared with
the corresponding picture fixed in the *Almagest*. As a measure of deviation
for this comparison, we will use the mean deviation of angular distances:

$$(1) \qquad \bar{\Delta}_i(t) = \frac{1}{N} \sum_{j=1}^{N} \left| \rho_{\text{real}}(S_i, O_j, t) - \rho_{\text{Alm}}(S_i, O_j) \right|$$

where N is the number of reference stars, $\rho_{\text{real}}(S_i, O_j, t)$ is the distance between the star S_i and the jth reference star O_j, and $\rho_{\text{Alm}}(S_i, O_j)$ is the distance
between the positions of the two stars in the *Almagest*. We will call the mo-
ment t at which $\Delta_i(t)$ attains the minimum value the *individual date from the
ith star*. If for most fast stars of the catalog the individual dates t_i concentrated
in a small temporal interval, then we should expect that the true date of compilation of the catalog is somewhere inside, or near this interval. *Regrettably,
this is not the case.*

3.   We have studied the behavior of deviations $\Delta_i(t)$ for eight stars,
Capella (222), Arcturus (110), Aquila (= Altair, 288), Denebola (488), Regulus (469), Sirius (818), Procyon (848), and Caph (189); in parentheses we give
the Baily number of the star in the *Almagest*). We deliberately chose "famous"
and brightest stars among fast stars of the *Almagest*, and abandoned fainter

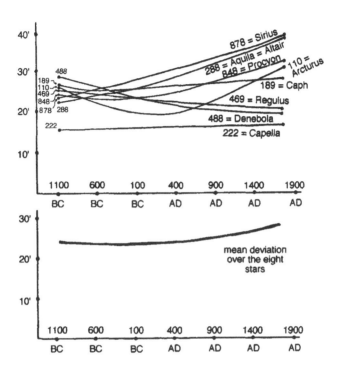

Figure 3.13. Latitudinal deviation $\bar{\Delta}_i(t)$ dependence on the a priori date t for the eight fast named stars of the *Almagest* (upper) and the graph of averaged deviation (lower).

stars, because, as we have noted above, the coordinates of faint stars in the *Almagest* are very inaccurate, so the inclusion of faint stars will only add to the scatter of individual dates. Figure 3.13 displays the graphs of deviations $\Delta_i(t)$ as functions of t. We also display the graph of mean deviation over all the eight stars; the mean deviation is almost constant all over the interval 1100 BC–1900 AD (bold line in Figure 3.13).

4. **Conclusions.** The abandonment of Newcomb's theory has led to no concentration of individual dates about any particular date. Hence, the wide scatter of dates is due to characteristics of coordinates given in the catalog (to their low accuracy, the presence of a significant systematic error, non-homogeneity of the catalog, etc.) rather than to the method of recalculation of coordinates. In the next section we analyze the stellar positions given in the *Almagest* and the general structure of the catalog, in order to reveal the characteristics of this kind.

4. Conclusions and directions of further study. Description of the method. Plan of the following chapters

1. In Sections 1–3 of this chapter we have exposed several attempts to date the *Almagest* from the numeric data contained in its star catalog; all the attempts ended in failure. We considered them for the following reasons. First, now the reader can estimate the difficulties that arise in dating the catalog "intrinsically" (from the data it contains). Second, we wanted to justify choice of the problems to be discussed in the sequel.

We should note that an attempt to date the *Almagest* from the numeric data it contains was made earlier [27,30]; an analysis of these works shows that the attempt is incorrect. Regrettably, the authors of Refs. 27 and 30, not specialists in mathematical statistics, misinterpreted the results of their computations and came to an unjustified conclusion about the date of compilation of the catalog. A detailed analysis of Refs. 27 and 30 is given in Ref. 31.

The main conclusion at this step is that dating the *Almagest* requires a thorough preliminary analysis of the catalog, touching upon the following questions.

1) *The problem of identification of stars in the Almagest with the stars in modern catalogs.* In Section 1 we have shown that the problem does not always find a firm solution and that the solution may depend on a presupposed date for the catalog. Therefore, before dating the catalog, it is necessary to reveal and exclude from further treatment all doubtfully or ambiguously identified stars.

2) *Analysis of the nature of possible errors occurring in the catalog.* The magnitudes of errors in coordinates of stars typical for the *Almagest* may lead to the conclusion that no nontrivial information about the date can be deduced from these data. However, there is a possibility to cope with this difficulty if we manage to isolate the *systematic component* of the errors. This will let us compensate for it, thus improving accuracy of the catalog, which probably will lead to a refinement of the estimate of the date of the catalog.

3) *Analysis of accuracy of the catalog reached in various collections of stars.* The aim of this analysis is to distinguish in the catalog groups of stars the coordinates of which were measured by Ptolemy to some guaranteed accuracy Δ. As soon as we distinguish such a group, it provides a set of possible dates for the catalog, namely, the dates when the real positions of stars were within Δ from the ones given in the catalog. If the ensuing set (the interval) is considerably smaller than the a priori fixed historical interval, this provides nontrivial information about the real date of compilation of the catalog. We use this idea in Chapters 5–7.

We will discuss here the three problems briefly; a more detailed analysis is carried out in subsequent chapters of this book.

2. The overwhelming majority of stars in the *Almagest* are unambiguously identified with stars in modern catalogs. Nevertheless, in Chapter 4 we carry out the identification anew, in order to be able to select stars for further treatment; the identification in most cases agrees with the one in Ref. 22, but we detect several stars whose identification with the stars in the *Almagest* depends on the choice of the epoch t (for example, o^2 Eri and μ Cas); in Ref. 27 these stars are identified under the presumption that the catalog of the *Almagest* was compiled about the 1st century AD. It is obviously senseless to use these stars as a base for dating the catalog, so we exclude them from further treatment. Note that in Ref. 22 the identifications of o^2 Eri and μ Cas are also stated as doubtful.

After identification, we obtain the table T the lines of which contain the following data about each identified pair of stars: (1) the Baily number of the star in the *Almagest*, (2) the direct ascent α_i and the declination δ_i of the star from the modern catalog at $t = 0$, (3) the components of velocity of proper motion of the star, (4) the longitude l_i and the latitude b_i of the corresponding star in the *Almagest*.

Denote by $\alpha_i(t)$ and $\delta_i(t)$ the equatorial and by $L_i(t)$ and $B_i(t)$ the ecliptic coordinates of the ith star in the computed catalog $K(t)$ for the year t. The problem is to find the value of t for which the set of coordinates $V(t) = (L_i(t), B_i(t))_{i \geq 1}$ or the set $W_i(t) = (\alpha_i(t), \delta_i(t))_{i \geq 1}$ agrees best with the set $V_A = (l_i, b_i)_{i \geq 1}$.

3. Above, we have shown that a straightforward comparison of true positions of stars with the ones given in the catalog does not lead to a reliable date for the catalog, due to the errors it contains. Therefore only the account of all errors may bring about a reliable date.

We will distinguish three types of errors, *group errors*, *random errors*, and *outlies*.

Group errors are the distortions of data that arise in measurements or recalculations and result in a displacement of a group of stars as a whole.

Random errors are the errors in coordinates of individual stars; mostly, they are introduced by errors of measurement within the scale division of the measuring instrument. Random errors displace each star by a random variable, whose mean value is zero.

Outlies originate in unexpected or unknown to the observer circumstances (refraction, copying errors, etc.). They affect coordinates of individual stars and usually have greater values than the scale division of the measuring instrument. The errors of this type are sufficiently sparse.

Our main goal is to find and compensate for the group errors. The way to do this is discussed in Chapter 5, where we give formulas for group errors and describe how to estimate the accuracy of the ensuing values. That accuracy problems are crucial in the problem, we have already had a chance to see.

The errors of the *Almagest* proper are treated in Chapter 6. *It turns out that the coordinates of stars in the Almagest carry considerable group errors that displace certain stellar configurations as a whole.* The values of the errors for different groups of stars may differ; this is where the term comes from. In fact, we will see that in comparatively large domains of the sky, the group errors for comparatively small constellations may coincide with each other and with the common group (systematic) error for the domain.

Each displacement corresponding to a group error may be described in terms of three parameters; we choose the following parameters (basic errors; see Figure 1.1).

1) *The error τ in the position of the spring equinoctial point $Q(t_A)$ made by the observer in the year of observation t_A, in the direction of the ecliptic.* In other words, τ is the projection on the ecliptic of the displacement of the position of the equinoctial point used in the catalog from its true position.

2) *The error β in the position of the point $Q(t_A)$ in the meridian direction* (the projection of the displacement on the ecliptic meridian).

3) *The error γ in the angle ε between the ecliptic and the meridian.* Any measurement of ecliptic coordinates requires a previous measurement of the angle ε. An error γ in this angle leads to the turn of the ecliptic used in the catalog through γ in relation to the true ecliptic.

The possibility of group errors has been discussed by many researchers[22,1,4]; here we will only mention their possible causes.

The error τ could arise from an inaccurate determination of the spring solstitial point. Another cause for this error is a reduction of the catalog by the observer or a later researcher to a date different from the date of original observations. Sometimes such a reduction was probably done for methodical reasons (say, an urge to reduce the catalog to a round date), and sometimes for disguising the real date of observations and attributing the catalog to a different epoch[1]; sometimes this was a consequence of a change of zero point for longitudes: generally, ancient astronomers could use various points for counting longitudes off. Any variation of the zero point led to adding a constant to all ecliptic longitudes. Clearly, τ *does not affect latitudes.* This is one of the reasons why latitudes are more reliable data than longitudes, and this is why we mainly work with latitudes, using information about longitudes as auxiliary. Consequently, *latitudinal deviations only require two parameters to determine group errors* (we use β and γ).

What can we say about β and γ? Since the equatorial coordinates of stars may be determined from immediate observations sufficiently easily and accurately[32], it should be expected that for an accurate observer, the error β must not be large. The error γ is of entirely different nature. An estimate of position of the ecliptic is a result of fairly complicated calculations or of nontrivial measurements (see Chapter 1), so γ may be substantially greater

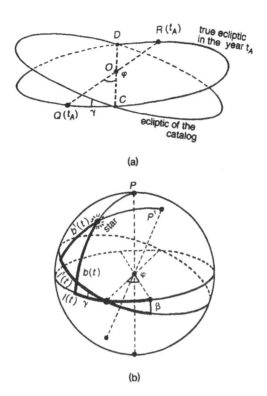

Figure 3.14. (a) Parametrization of systematic error in terms of γ and φ. (b) Geometric sense of the parameters γ, β and φ.

than β. Some indications that the systematic error γ exists can be found in Refs. 22 and 4; moreover, some authors estimate it at about 20′. Our calculations in Chapter 6 confirm this estimate.

Sometimes, we will use parameters φ and γ, more convenient for computations than β and γ. Figure 3.14 explains the sense of these variables. From the point of view of latitudinal deviations, the error reduces to the error in the position of the ecliptic plane; we will call the (inexact) position of the ecliptic used in the catalog the *ecliptic of the catalog*. The position of the ecliptic of the catalog in relation to the true ecliptic in the year t_A of compilation of the catalog may be described in terms of the angle φ the equinoctial axis QR (of the year t_A) makes with the axis CD where the planes of the true ecliptic and of the ecliptic of the catalog meet, and the angle γ between the two planes. We will use the angles φ and γ as parameters for the group error.

Generally, the compiler of the catalog could make different group errors in various domains of the sky, say, because of resetting the instrument or change of the vantage point, etc.

In Chapter 2 we have distinguished seven domains of the sky (Figure 2.2), differing in reliability of measurements. In Chapter 6 we will see that they also differ in group errors and in accuracy of measurement.

4. We do not know the moment t_A of compilation of the *Almagest*, so for each moment t we compute the values $\varphi(t)$ and $\gamma(t)$; the method of computation is a combination of the method of least squares with the regression problem on a sphere. The accuracy of the method is discussed in Chapter 5. As a result, we obtain graphs of two functions, $\gamma_{stat}(t)$ and $\varphi_{stat}(t)$ (Figure 3.15), that ensue from statistical processing of coordinates of stars in various domains of the sky. The subscript stat indicates that the functions come from statistical processing; the two functions represent regular components of the errors present in *large domains* of the sky under the assumption that the true date of the catalog is t. We call such errors *systematic*. In *large* domains of the sky, consisting of several constellations, statistically computed systematic errors are mean values of group errors characteristic for individual constellations. Of course, *only in the case that all group errors are equal to each other do they coincide with the systematic error*, so only in this case we do not distinguish group errors and systematic errors. About the values $\gamma_{stat}(t)$ and $\varphi_{stat}(t)$, we will define the so called *confidence intervals* I_γ and I_φ, the necessity of which is due to the statistical nature of our computations, whose certainty is not 100%. Therefore, we only can say that the real values of parameters lie with a certain probability in a neighborhood of the values φ_{stat} and γ_{stat} we have found. The construction of confidence intervals is given in Chapter 5, and in Chapter 6 we expose the results concerning the *Almagest*.

We applied the above scheme to each of the seven domains of the sky, and found the values of systematic errors as well as the values of "residual" mean square latitudinal deviations; as a result, we found out that the domains A and *Zod* A are measured best (see Chapter 6 and Table 2.3), which, by the way, contain a majority of named stars of the *Almagest*. We found out also (see Chapter 6) that after the systematic error is compensated for, *more than half* of the stars in the domain A have the latitudinal deviation within $10'$; for *Zod* A the rate of such stars is still higher, 63.7%. Thus, *the claimed accuracy of the catalog* ($10'$) *acquires confirmation for the well-measured part of the sky*.

The next question that arises is the one on the nature of the parameters γ_{stat} and φ_{stat}, namely, whether their values coincide with the real group errors for all stars (or, say, for the domain A). The fact is that individual constellations may have different group errors, the mean of which is the value we have found. We have considered all zodiacal constellations and neighborhoods of most named stars to see that the value $\gamma_{stat}^{Zod\ A} \approx \gamma_{stat}^{A}$ is equal to the group errors at least for all constellations in A. In other words, $\gamma_{stat}^{Zod\ A}$ *is to be treated*

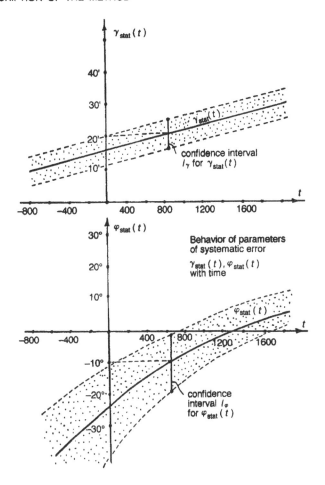

Figure 3.15. Dependence of the parameters of the systematic error γ_{stat} and φ_{stat} on the a priori date, and the confidence zones for these estimates.

as a regular component that effects all stars of the well-measured domain A, which in addition contains a majority of named stars. This statement is no longer true for $\varphi_{stat}^{Zod\ A}$. It is interesting that this may serve as an argument for the statement that an armillary sphere was used for observations for the *Almagest* (see Chapter 6).

5. Compensation for the systematic error improved the accuracy of the catalog (in *Zod* A, from $17'.7$ to $12'.6$), which enhanced the possibilities of dating.

As we have already mentioned, dating requires considering well-measured fast stars. We already know that the claimed accuracy of the catalog of the *Almagest* is *really reached for a considerable part of the stars in the catalog*. Are there any stars for which the accuracy is reached *for sure*?

Usually, measuring stellar positions, the observer uses a system of reference points (stars)[33]. The reference stars were used by all ancient and medieval astronomers and are used in modern astronomy. For example, Tycho Brahe based his observations on a system of 21 reference stars[34]. Modern system of reference points involves several thousands stars (described in the so-called *fundamental catalogs*; see, for example, the catalog $FK4$[29]). We can derive from the structure of the catalog of the *Almagest* that Regulus and Spica were used as reference points: special sections of the *Almagest* are devoted to measuring their positions.

We propose the hypothesis that *if the claimed accuracy is confirmed, then the accuracy is attained on the set of famous stars in the catalog*. It seems natural to consider as "famous" the *named* stars, that is, the ones that have proper names in the catalog. There are twelve such stars, and they really form a very convenient frame in the visible part of the sky. All these stars are bright and are easily distinguishable among nearby stars. What is important, some of them have high velocities of proper motion (Arcturus, Procyon, and Sirius), and some others also move notably in the celestial sphere (Regulus, Capella, Antares, and Aquila).

We excluded at once from our further treatment two of the twelve stars, Canopus and Vindemiatrix. Refraction distorted greatly the coordinates of Canopus, so this star is an outlie; as for Vindemiatrix, Ptolemy's original coordinates of this star are simply unknown (see Chapter 2). Two more stars, Sirius and Aquila, were excluded because the systematic errors for their close environment differs from the one for the environments of other stars (as follows from our analysis), but we cannot determine these errors. Thus, our dating is based on eight named stars.

6.   The above hypothesis implies the statement that at the year of compilation of the catalog t_A, latitudinal deviation of *all* the eight stars was below 10′.

On the other hand, we know that in the year t_A the component γ of the systematic error of the catalog lies in the confidence interval I_γ, that contains the value $\gamma_{stat}(t_A)$. Hence a natural dating procedure ensues.

Let us consider, at some fixed t, the confidence interval I_γ, $\gamma_{stat}(t)$ and distinguish its subset S_t of all values of γ such that after the compensation for the systematic error γ in the component, the latitudinal deviations of all the eight named stars do not exceed 10′ — see Figure 3.16. Of course, for some values of t the set S_t is empty; *let us find all values of t for which the set S_t is not empty*. The values of t thus found form the *interval of admissible dates* for the catalog, because it contains exactly those values of t at which the latitudes of all the eight stars could be measured with the accuracy 10′.

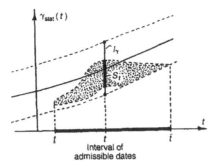

Figure 3.16. Statistical dating procedure. Intersection of the confidence zone with the domain where the minimax latitudinal error does not exceed 10′. The bold interval in the time axis is the resulting interval of admissible dates.

We note by $\bar{t}$ and $\underline{t}$ the lower and the upper boundaries of the interval. We call this procedure *statistical*, because it is based on the magnitudes $\gamma_{stat}(t_A)$, found from statistical considerations. A detailed description of this approach is exposed in Chapter 7, where we also present the results of application to the catalog of the *Almagest*. The resulting dating interval has the lower bound 600 AD and the upper bound 1300 AD. Although it is 700 years long (because of low accuracy of the *Almagest*), this interval is far from the date traditionally attributed to the *Almagest*.

7.   The confidence intervals used in the statistical approach depend on an arbitrarily fixed parameter, the *confidence level* (the probability with which the results are true). Therefore, generally speaking, we could discuss the dependence of the interval on the confidence level. Similarly, the conjecture that the group error for the eight stars is equal to the systematic error for *Zod A* is of statistical nature, hence may be false with some probability. Therefore, the question arises: How can the resulting interval expand if we alter the confidence level? It is supposed to give a "geometrical" answer to this question. Let us again fix a moment of time t as a "candidate" for the date. Let us find the set D_t of all values of γ whose turn through the true ecliptic of the year t renders the latitudinal deviations of the eight named stars less than or equal to 10′ (see Figure 3.17). Clearly, $D_t \supset S_t$ for all t. Hence, this approach gives *all* possible values of t at which it is possible to turn the ecliptic so that the latitudes of all the eight named stars differ from their values given in the *Almagest* by less than 10′. It is far from obvious that *the ensuing "maximum possible" interval coincides with the one obtained from the statistical procedure* (see Chapter 7).

In Chapter 7 we also show *stability* of the dating procedure with respect to variation of the starting assumptions (the claimed accuracy and nonlinear

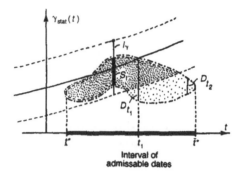

Figure 3.17. Geometric dating procedure. The projection of the domain of values of parameters where minimax latitudinal deviation does not exceed 10′ on the time axis is the resulting interval of admissible dates (bold interval in the time axis).

distortions of observational instruments). The method has been tested on artificially created catalogs and on several reliably dated catalogs. The results of the testing are exposed in Chapter 9.

8.   Thus, the following steps are to be done to come to the date of the catalog:

1) Reveal the doubtfully identified stars in the catalog and exclude these stars from further treatment. Also, find and exclude the stars whose coordinates are given in the catalog with very large errors(the outlies).

2) Find and compensate for systematic errors of the catalog on the whole or of large parts of it. This improves the accuracy of the catalog.

3) Confirm (or refute) the claim of the compiler that the accuracy of the catalog is 10′ by finding sufficiently many stars that have latitudinal deviations below 10′ (after compensation for the group error).

If the claimed accuracy is not confirmed, we should consider as such the so-called *record accuracy*, that is, the one to which about 50% of the stars are measured.

4) Determine the *informative kernel* of the catalog, the set of stars at which the claimed accuracy is surely attained.

5) Determine the dating interval as the set of all moments of time at which the latitudinal deviations of stars in the informative kernel do not exceed the claimed accuracy of the catalog (after compensation for the "statistical" or "geometrical" group error).

We realize these steps in Chapters 4–7.

Part Two
Dating Catalogs

Chapter 4
Who is Who?

1. Preliminary remarks

We have already remarked that the solution of the dating problem may be affected by the chosen identification of stars in the *Almagest* with stars in modern catalogs. The identification problem is not trivial and is not uniquely solved in all cases. Recall that the catalog of the *Almagest* comprises 1025 stars, divided into constellations and informatas. Clearly, this partition is rather arbitrary. Only twelve stars have proper names (Arcturus, Aquila, Antares, Vindemiatrix, Aselli, Procyon, Regulus, Spica, Vega (Lyra), Capella, Canopus, Sirius (in fact, the compiler uses the word *Canis* instead of *Sirius*)). The rest of the stars have no names and are endowed with descriptions like "the star in the middle of the neck", "the star on the end of the tail", "the brighter of the two stars in the left knee", etc. Of course, such descriptions are insufficient for a firm identification of the star described with a star in a modern catalog. Many researchers of the *Almagest* carried out the identification based on comparison of stellar positions given in the *Almagest* with the positions of now known stars; some results may be found in the work of Peters and Knobel [22], where a modern star is assigned to each star in the *Almagest*. The work also contains a table of variants of identification suggested by various researchers. The identifiers believed in the hypothesis that the catalog had been compiled about the 1st century AD, and in some cases this influenced the identification. The fact is that some stars that have high velocity of proper motion, alter their position with time, approaching positions of various stars

in the *Almagest* in various epochs. Such stars should not be used in dating
the catalog, since the date they imply depends on the particular identification,
which in its turn depends on the presupposed date. Furthermore, some fast
stars may have no companion in the *Almagest*, because most fast stars are
faint (fourth to sixth star magnitude), and not all faint stars are reflected in
the *Almagest*. We also cannot exclude the situation when more than one star
matches a description given in the *Almagest*. We need to detect all such cases
to avoid basing our dating procedure on ambiguities.

We have never had doubts in thoroughness of the identification carried
out in Ref. 22 (confirmed by our computation). What we mean here is the
mistakes generated by the presupposed attributing of the catalog to the 1st
century AD. In order to exclude all doubts, we have carried out the process
of identifying fast stars anew.

2. The method for identification

As we have noted, the question *who is who?* is essential mainly for fast stars,
most of which are *faint*. The identification of *named and bright* stars carried
out in Ref. 22 raises no doubts. The faster a star moves, the more accurate is
the date for the catalog it provides (of course, if the star is identified exactly
and firmly with a star in the catalog). We took for identification 82 fast stars
in the modern catalog[21] (those that have velocity at least $0\overset{''}{.}5$ per year in at
least one of the coordinates in the equatorial coordinate system of 1900 AD).
The list of such stars is in Table Ap. 1 in the Appendix, where we also give
the most important characteristics of the stars: their equatorial coordinates
in 1900 ($t = 0$ in our notation) and the velocities of proper motion. Using
these data and the formulas for passage to ecliptic coordinates (see Chapter 1)
and to account for proper motion, we find the ecliptic coordinates $L_i(t)$ and
$B_i(t)$ of the ith star in the year t ($1 \leq i \leq 82$). For each of the 82 fast
"modern" stars we consider its ε-neighborhood (that is, the circle of radius
ε) in the celestial sphere (see Figure 4.1). Further, for a fixed t (we carry
out the computation for all values of t in the interval from 0 to 30, which
corresponds to the temporal interval 1100 BC–1900 AD) we compute the
angular distance $\xi(A, i, t)$ between the coordinates (l_A, b_A) of the star A in
the catalog of the *Almagest* and the coordinates $(L_i(t), B_i(t))$ of the ith "fast
modern" star computed for the moment t. If $\xi(A, i, t) < \varepsilon$, then we conclude
that at the date t, the star A of the *Almagest* identifies with the ith fast star of
the modern catalog. Thus, the identification occurs when the ε-neighborhood
of the ith star captures a star from the *Almagest* in a temporal interval $[t_*, t^*]$.
Of course, it may happen that more than one star of the *Almagest* gets into
the ε-neighborhood of a modern star, either simultaneously or at different
moments of time. Also, it may occur that no position of a star of the *Almagest*
ever gets into the ε-neighborhood.

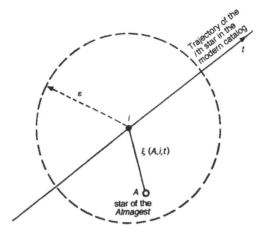

Figure 4.1. Identification of the stars in the *Almagest* with the stars of the real sky. The capture of a star of the *Almagest* by the ε-neighborhood of trajectory of true motion of a real star.

It is obvious from the above description that the method is rather rough; in particular, we should choose the radius ε several times as large as the accuracy of the catalog under study, to be sure of reliability of the identification. As we will see, the ensuing identification is practically independent of the choice of ε, which confirms the stability of the method.

3. The resulting identification of "modern" stars with the stars in the *Almagest*

The accuracy claimed by the compiler of the catalog of the *Almagest* is 10' (separately in latitudes and in longitudes). This means that the (claimed) accuracy in angular distances is about 14' ($\sqrt{2}$ times the accuracy of measurement of each coordinate). However, the claimed accuracy, generally speaking, is the *record accuracy*, only reached at the stars measured best (say, the named stars); as for the *real* accuracy, it may be several times worse. A detailed discussion of accuracy problems is presented in the next two chapters. Here we merely choose ε a few times as large as the claimed accuracy 14'; namely, we took the values $\varepsilon = 0°5$, $\varepsilon = 1°$, $\varepsilon = 1°5$ and $\varepsilon = 2°$. In Table 4.1 we expose the results of identification of fast stars in the temporal interval $0 \leq t \leq 30$ (1100 BC–1900 AD). Of the 82 fast stars in Table Ap. 2, we only give here the data for those stars whose ε-neighborhoods capture at least one star of the *Almagest* at some values of t and ε. Each line of the table exposes a pair of a "fast modern" star, whose number is given in Ref. 21, and

Table 4.1

Number in Ref. 21	Baily's number	$\varepsilon =$			
		0°5	1°0	1°5	2°0
21	189	[20,30]	[0,30]	[0,30]	[0,30]
219	180	—	[0'30]	[0,30]	[0,30]
321	185	—	[6,27]	[0,30]	[0,30]
509	723	[4,30]	[0,30]	[0,30]	[0,30]
660	360	[8,30]	[8,30]	[8,30]	[8,30]
	361	[0,7]	[0,7]	[0,7]	[0,7]
753	716	—	[10,30]	[2,30]	[0,30]
937	196	[27,30]	[0,30]	[0,30]	[0,30]
1136	783	[0,13]	[0,30]	[0,30]	[0,30]
1325	778	[29,30]	[29,30]	[29,30]	[29,30]
	779	[19,25]	[14,28]	[12,28]	[12,28]
	780	—	[0,8]	[0,11]	[0,11]
1614	775	—	—	[0,30]	[0,30]
2491	818	[8,30]	[0,30]	[0,30]	[0,30]
1943	848	[0,17]	[0,30]	[0,30]	[0,30]
2990	425	[0,30]	[0,30]	[0,30]	[0,30]
2998	882	[0,30]	[0,30]	[0,30]	[0,30]
4375	32	[0,3]	[0,30]	[0,30]	[0,30]
4414	486	[0,30]	[0,30]	[0,30]	[0,30]
4540	501	—	[14,30]	[0,30]	[0,30]
4657	732	—	—	[0,30]	[0,30]
5019	527	[8,30]	[0,30]	[0,30]	[0,30]
5188	935	—	—	[0,30]	[0,30]
5288	940	—	[0,21]	[0,30]	[0,30]
5340	110	[5,13]	[0,25]	[0,30]	[0,30]
5460	969	—	—	—	[0,30]
5699	979	[0,25]	[0,30]	[0,30]	[0,30]
5933	265	—	[8,30]	[0,30]	[0,30]
6241	557	[0,30]	[0,30]	[0,30]	[0,30]
6401	247	[17,30]	[0,30]	[0,30]	[0,30]
6623	125	—	[0,30]	[0,30]	[0,30]
6752	261	—	[4,30]	[0,30]	[0,30]
6869	279	[0,28]	[0,30]	[0,30]	[0,30]
7957	79	—	[0,22]	[0,30]	[0,30]
8085	169	—	—	[22,30]	[20,30]
8697	327	—	—	[0,7]	[0,7]
	328	[28,30]	[8,30]	[8,30]	[8,30]
8775	317	[0,30]	[0,30]	[0,30]	[0,30]
8969	678	[0,30]	[0,30]	[0,30]	[0,30]

a star of the *Almagest*, whose Baily number is given. We write "—" if the
ε-neighborhood of the star in the first column never captures the star in the
second column at the given ε. For example, at $\varepsilon = 0°5$, the ε-neighborhood
of the star 1325 in Ref. 21 never captures the star with Baily's number 780.
If the star of Ref. 21 identifies with a single star in the *Almagest*, we write

in the line the Baily number of this star and the temporal intervals in which the identification occurs (for various values of ε). For example, the star 21 in Ref. 21 (11 β Cas) identifies with the star number 189 at $28 \leq t \leq 30$ for $\varepsilon = 0°5$ and at $0 \leq t \leq 30$ for the rest of the values of ε. If the star i in the first column identifies with several stars, we indicate them all, and in the corresponding lines write down the intervals in which the star of the *Almagest* is nearer to the star i than the rest. For example, the star 1325 in Ref. 21 (40 o^2 Eri) identifies in various temporal intervals with the stars of the *Almagest* having Baily numbers 778, 779 and 780. The column corresponding to $\varepsilon = 1°5$ tells that at $0 \leq t \leq 10$ the star of the *Almagest* nearest to the star 1325 is Baily's 780 (although, say, at $t = 10$ the distance between 1325 of Ref. 21 and Baily's 779 is also less than $1°5$).

The sense of the above calculations is the following. Suppose the catalog was compiled in the year t. Then the most natural candidate for being the star that has Baily number A is the "modern" star that identifies with A in Table 4.1.

It is obvious from Table 4.1 that the results of the identification are almost insensitive to the choice of ε. In fact, this choice is fairly arbitrary and is due to informal considerations. First, ε should be of the same order with the accuracy of the catalog (otherwise we will have to consider identifications that have nothing in common with reality). Second, it should be sufficiently large in order that we have any identifications at all and that the errors of the catalog could not affect the result essentially. Third, ε should not be so large as to make the results of identifications ambiguous.

It follows from Table 4.1 that we can identify 36 of the 82 fast stars. The ensuing identifications do not contradict the ones in Ref. 22; moreover, most of them coincide with the known ones. A dramatic exception is the star 1325, o^2 Eri. In Ref. 22 the star is marked as doubtfully identified. We have come to several possible identifications, depending on the date of compilation of the catalog. Taking into account its faintness, we should treat the stars with Baily's numbers 778, 779 and 780 as doubtfully identified and exclude them from further consideration.

The results represented in Table 4.1 also show that reidentifications are exceptions rather than the rule, which may be explained by slowness of the overwhelming majority of stars, as well as by the sparseness of the stars of the *Almagest* in the celestial sphere. In our further study, we confine ourselves to the stars that have no ambiguities in identification. Therefore we will use their Baily numbers, not referring to their numbers in Ref. 21. In some cases we will also use proper names of the stars.

Although Table 4.1 demonstrates a relation between the "accuracy of identification" ε and temporal intervals, no reliable information about the date may be elicited hereof. The reasons for that are discussed in details in Chapter 3; nevertheless, we will repeat them briefly here. If we delete from the list of all fast stars the ambiguously identifiable ones and choose ε as the minimum value at which *all* temporal intervals meet, this value might be treated as the real accuracy of measurement of fast stars, and the point where the intervals first meet as the true date of compilation of the catalog. However, as follows from Table 4.1, this value of ε is too large: even the fastest stars cover such distances in *thousands of years*. The ensuing date turns out to be extremely unstable; in particular, it depends heavily on the collection of stars it is based on — an addition or deletion of a single star may bring about a large variation of the date. This is why we have required classification of stars from the point of view of accuracy of measurement as a necessary step of the dating procedure.

4. Conclusions

1. An overwhelming majority of stars in the catalog of the *Almagest* have been identified correctly in previous studies.[22]

2. 36 of 82 fast stars in the modern catalog[21] admit identification with stars in the *Almagest* (see Table 4.1).

3. The following stars in Table 4.1 admit various identifications:

a) o^2 Eri (40 o^2 Eri, 1325) may be identified with the star 778 of the *Almagest* in the interval 1100 BC–800 BC, with the star 779 of the *Almagest* in the interval 700 BC–800 AD, and with the star 780 of the *Almagest* in the interval 900 AD–1900 AD.

b) The star 660 of Ref. 21 may be identified with the star 361 of the *Almagest* in the interval 1100 BC–1800 AD, and with the star 360 of the *Almagest* in the interval 1800 AD–1900 AD.

c) The star 8697 of Ref. 21 may be identified with the star 327 of the *Almagest* in the interval 1200 AD–1900 AD and with the star 328 of the *Almagest* in the interval 1100 BC–1200 AD.

Chapter **5**

Analysis
of Systematic Errors
in Stellar Configurations

1. Classification of latitudinal errors

From this chapter on, we assume that we deal with a catalog all of whose stars admit a well-defined identification with the stars in a modern catalog. The variable i ranges over the numbers of stars in the catalog; l_i and b_i stand for the ecliptic longitude and latitude of the ith star in the catalog, and $L_i(t)$ and $B_i(t)$ for the true ecliptic longitude and latitude of the star in the year t. Recall that we count the time t in centuries from 1900 to the past; thus, $t = 3.15$ corresponds to the year $1900 - 3.15 \cdot 100 = 1585$ (AD), and $t = 22.0$ to the year $1900 - 22.0 \cdot 100 = -300$ ($= 300$ BC). Let t be the (unknown) year of compilation of the *Almagest*. Denote by L_i^A and B_i^A the true longitude and latitude of the ith star in the year t_A: $L_i^A = L_i(t_A)$, $B_i^A = B_i(t_A)$, and let $\Delta B(t) = B_i(t) - b_i(t)$ be the *latitudinal deviation* (for the year t); we will use $\Delta B_i(t)$ as a measure of the error in the latitude of the ith star in the catalog *under the condition that the catalog was compiled in the year t.* Naturally, $\Delta B_i(t_A) = \Delta B_i^A$ is the true error in latitude.

We confine ourselves to considering latitudinal errors, for the reasons explained in Chapter 3.

We classify the errors occurring in the catalog into four types: the *outlies*, the *group errors*, the *systematic errors*, and the *random errors*. We call *outlies* the gross errors in coordinates; they are usually easy to detect. We exclude such stars from our further consideration. *Group errors* are the ones introduced in the coordinates of a group of stars in a similar way. A typical example is the error introduced by an erroneous determination of position of the ecliptic in the celestial sphere. The group errors that are common for the whole catalog, or in a large part of the catalog, we call *systematic*. The systematic errors may be detected and compensated for. *Random errors* are the ones that admit no compensation in principle (for example, random instrumental errors that have no regular component).

Note that *we only classify errors, but not their causes*. In particular, the outlies may be caused by mistakes of copyists, refraction, etc. The group errors may be introduced by faulty instruments, astrometric errors, etc. Similarly, the causes of random errors are various; it is apparently impossible to enlist, not to say detect all the causes. We need not do that; the approach we propose uses the values of the errors, but not their causes.

A lot of preliminary work for detecting outlies has been done in several well-known investigations of the *Almagest* (see, for example, Ref. 22), where the values of latitudinal deviations are given for all stars). We classified as outlies the stars whose latitudinal deviations exceed 1°. Note by the way, that a variation of this critical level practically does not affect our conclusions, because the outlies are relatively few. Therefore, weeding out the outlies constitutes no difficult problem, and in the sequel we assume that no outlies are left. The aim of the methods we present below is to compensate for group errors (thus improving the accuracy of the catalog) and to try to date the catalog with only the random errors left.

2. Parametrization of group and systematic errors

Consider a collection of stars (a constellation or a group of constellations). We define the *group error* in coordinates (latitudes) of these stars as the error introduced by a translation along the celestial sphere of the collection of stars as a whole. Consequently, *any subcollection of the configuration is translated in the celestial sphere as a whole, by the same angle as the whole collection* (we stress this, because we will use this in the sequel). A translation along the sphere has three degrees of freedom, so three parameters are needed to determine it. Let us introduce the three parameters.

Figure 5.1 shows the picture. In the celestial sphere with the center O, the true position of the ecliptic in the year t_A is depicted, in which the spring and the fall equinoctial points Q and R are marked. The point E depicts the position of a star. As we have noted, the group errors in latitudes (for a fixed group of stars) may be considered as a consequence of an incorrect determination of the pole of the ecliptic; that is, some point P_A was assumed to be the pole instead of the point P. A perturbed ecliptic corresponds to this choice of P_A, which we call *the ecliptic of the catalog*. Its position may be defined in terms of two parameters, the angle γ between the lines OP and OP_A (or, which is the same, the angle between the planes of the true ecliptic

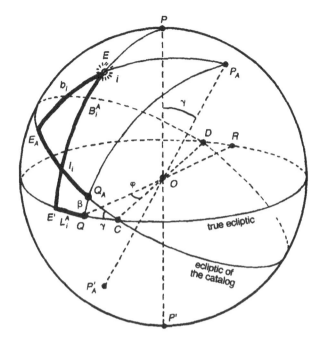

Figure 5.1. Parameters that determine systematic error (the error in the position of the ecliptic of the catalog).

and of the ecliptic of the catalog) and the angle φ between the equinoctial axis RQ and the line CD where the plane of the true ecliptic and the plane of the ecliptic of the catalog intersect. This parametrization is convenient for calculations.

We will also sometimes use the angle β, defined as follows (see Figure 5.1). The turn of the ecliptic may be decomposed into two turns, the turn about the equinoctial axis RQ through the angle γ and the turn about the axis that lies in the ecliptic plane perpendicular to RQ through the angle β. Thus, β is the angle subtended by the arc $Q_A Q$ of the large circle through the pole P_A and Q. The astronomic sense of the point Q_A is quite obvious: it is the spring equinoctial point of the catalog. Clearly, the values of γ and φ determine uniquely the values of γ and β and vice versa. The particular relation between them may be found from the right spherical triangle CQ_AQ (the angle at the vertex Q_A is right, the angle at the vertex C is γ, and the arc CQ subtends the angle β); we get

(1) $$\sin \beta = \sin \gamma \cdot \sin \varphi$$

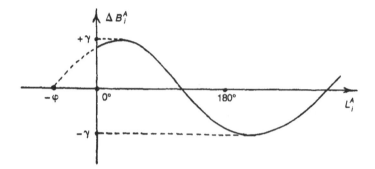

Figure 5.2.   Systematic latitudinal deviation dependence of longitude.

The third degree of freedom is the turn of the sphere about the axis $P_A P'_A$ (see Figure 5.1); it does not affect latitudes, so we do not consider it in our analysis. (Of course, any other parameters that determine turns of the sphere may be chosen.)

Let us now look how the systematic error determined by γ and φ distorts the coordinates of the ith star. The true longitude L_i^A and latitude B_i^A are equal, respectively, to the lengths of the arcs EE' and QE' (counted clockwise if looked at from the pole P). The distorted latitude b_i and longitude l_i are equal, respectively, to the lengths of the arcs EE_A and $Q_A E_A$. Note that the latitudes of the stars whose true latitude is more than the latitude of D and less than the latitude of C (see Figure 5.1) decrease, and the latitudes of the rest of the stars increase. Strictly speaking, this is not true for all stars, namely, it is wrong for the stars within γ from the poles P and P'; because of the smallness of distortions introduced by the compiler of the catalog (as we will see, γ is about 20'), there are practically no such stars.

Taking into account the smallness of γ, we may use the following approximate expression for the latitudinal deviation:

$$(2) \qquad\qquad \Delta B_i^A = \gamma \sin(L_i^A + \varphi)$$

Thus, the systematic error in latitudes may be depicted as the sine curve in Figure 5.2, which it is worthwhile to compare with the Peters' sine curve (see Ref. 22 or Figure 2.11).

The error of the formula (2) does not exceed 1' for the stars with $|b_A| \le 80°$. This error is immaterial for us, so below we will treat the formula as exact (to justify this, we exclude from further considerations all stars whose latitudes exceed 80°). Further in this chapter, we will speak of the systematic error, because the methods we expose work under the assumption that a sufficiently large collection of stars is considered. A check of the coincidence of the resulting systematic error with the group errors of individual constellations is a separate problem, which we deal with in the next chapter.

If we knew the year t_A when the catalog was compiled, we would be able to determine the parameters γ and φ from the following procedure:

1) Find the true coordinates B_i^A and L_i^A of all star in the collection.
2) Find the parameters γ and φ so that

$$(3) \qquad \sigma^2(\gamma, \varphi) \to \min$$

where

$$(4) \qquad \sigma^2(\gamma, \varphi) = \sum_i (B_i^A - b_i - \gamma \sin(L_i^A + \varphi))^2$$

Suppose for a moment that the catalog has no errors but systematic. Then (3) reduces to

$$(5) \qquad \sigma^2(\gamma, \varphi) = 0$$

However, we in fact do not know the year of compilation of the catalog, so we have to compute the systematic errors for all values of t in the interval $0 \le t \le 25$. Namely, for each t we first find the position of the true ecliptic and the equinoctial axis. Then, as in Figure 5.1, we introduce parameters $\gamma = \gamma(t), \varphi = \varphi(t)$ and $\beta = \beta(t)$ that determine the position of the ecliptic of the catalog in relation to the true ecliptic of the year t. The values $\gamma(t)$ and $\varphi(t)$ may be found from the condition

$$(6) \qquad \sigma^2(\gamma(t), \varphi(t), t) \to \min$$

where

$$(7) \qquad \sigma^2(\gamma, \varphi, t) = \sum_i (\Delta B_i(t) - \gamma \sin(L_i(t) + \varphi))^2$$

Again, if the catalog contained no errors but systematic, the relation (6) would reduce to the equation

$$(8) \qquad \sigma^2(\gamma, \varphi, t) = 0$$

which would have solutions at any t. So, it is impossible to find the date t_A from this equation; what we can find is the systematic error as a function of t. Naturally, the systematic error would vary because of oscillations of the ecliptic. That is why we speak of finding the systematic error of the catalog, but not of dating it.

Since the real catalog contains also random errors, the deviations $B_i(t) - b_i$ are random variables the values of which concentrate about the sine curve in Figure 5.2. If we assume that the errors in the catalog, except systematic, are random, say, normally distributed, then the problem of finding $\gamma(t)$ and $\varphi(t)$ reduces to the problem of finding parameters of regression, solving which we find statistical estimates for $\gamma(t)$ and $\varphi(t)$.

3. Method of least squares for the parameters γ and φ

In this section, we will find the solution $\gamma(t)$ and $\varphi(t)$ for the minimization problem (2.6)–(2.7). Since below we will consider various collections of stars, it is convenient to use *normed* magnitudes,

(1)
$$\sigma_0^2(\gamma, \varphi, t) = \frac{1}{N} \sum_{i=1}^{N} \sigma^2(\gamma, \varphi, t)$$

(2)
$$s_b(t) = \frac{1}{N} \sum_{i=1}^{N} \Delta B_i(t) \cdot \sin L_i(t)$$

(3)
$$c_b(t) = \frac{1}{N} \sum_{i=1}^{N} \Delta B_i(t) \cdot \cos L_i(t)$$

(4)
$$s_2(t) = \frac{1}{N} \sum_{i=1}^{N} \sin^2 L_i(t)$$

(5)
$$c_2(t) = \frac{1}{N} \sum_{i=1}^{N} \cos^2 L_i(t)$$

(6)
$$d(t) = \frac{1}{N} \sum_{i=1}^{N} \sin L_i(t) \cos L_i(t)$$

where N is the number of stars in the collection in question.

Note that all these magnitudes may be found for any year t from modern coordinates of stars and the coordinates of stars given in the catalog.

Obviously, the minimization problem (2.6) is equivalent to

(7)
$$\sigma_0^2(\gamma, \varphi, t) \to \min$$

in the sense that the parameters $\gamma(t)$ and $\varphi(t)$ determined by (7) are the same as the ones determined by (2.6).

Since, as we noted, the problem (7) makes sense only for large collections of stars, and since we will only study statistical properties of its solution, we will denote the values that satisfy (7) by $\gamma_{\text{stat}}(t)$ and $\varphi_{\text{stat}}(t)$.

The value

(8) $$\sigma_{\min}(t) = \sigma_0(\gamma_{\text{stat}}(t), \varphi_{\text{stat}}(t))$$

has an obvious physical sense: it is the residual mean square latitudinal deviation (over the collection of stars, at the moment t) that we obtain *after compensation for the systematic error* $\gamma_{\text{stat}}(t)$, $\varphi_{\text{stat}}(t)$. As we will see below, $\sigma_{\min}(t)$ in fact *does not depend on t* (for groups consisting of fixed stars); therefore we will use the denotation $\sigma_{\min}$. Note that *with the group error not compensated for*, the mean square latitudinal deviation at the moment t is equal to

(9) $$\sigma_{\text{init}} = \sigma_0(0, 0, t) = \sqrt{\frac{1}{N} \sum_{i=1}^{N} (\Delta B_i(t))^2}$$

This magnitude, generally speaking, depends on t. Thus, the difference

(10) $$\Delta\sigma(t) = \sigma_{\text{init}}(t) - \sigma_{\min}(t)$$

estimates the effect of compensation for the systematic error determined by $\gamma_{\text{stat}}(t)$ and $\varphi_{\text{stat}}(t)$.

Further, as we determine $\gamma_{\text{stat}}(t)$ and $\varphi_{\text{stat}}(t)$, we will assume t fixed, so we will omit t in the expressions (that is, we write L_i instead of $L_i(t)$, s_b instead of $s_b(t)$, etc.).

In order to find minimum in the relation (7), let us find partial derivatives of $\sigma_0^2(\gamma, \varphi, t)$ with respect to γ and φ and equate them to zero. Applying

(11) $$\sin(L_i + \varphi) = \sin L_i \cos \varphi + \cos L_i \sin \varphi$$

we come to the equations

(12) $$\begin{aligned} & s_b \cos \varphi + c_b \sin \varphi \\ & = \gamma (s_2 \cos^2 \varphi + 2d \cos \varphi \sin \varphi + c_2 \sin^2 \varphi) \end{aligned}$$

(13) $$\begin{aligned} & -c_b \cos \varphi + s_b \sin \varphi \\ & = \gamma (-d \cos^2 \varphi + (s_b - c_b) \cos \varphi \sin \varphi + d \sin^2 \varphi) \end{aligned}$$

Dividing (12) by (13), we get

(14) $\qquad \dfrac{s_b + c_b \cos\varphi \tan\varphi}{-c_b + s_b \tan\varphi} = \dfrac{s_2 + 2d\tan\varphi + c_2\tan^2\varphi}{-d + (s_2 - c_2)\tan\varphi + d\tan^2\varphi}$

whence

(15) $\qquad (1 + \tan^2\varphi)(c_b s_2 - s_b d) + (1 + \tan^2\varphi)\tan\varphi(c_b d - s_b c_2) = 0$

Solving (15), we find

(16) $\qquad\qquad\qquad \tan\varphi_{\text{stat}} = \dfrac{s_b d - c_b s_2}{c_b d - s_b c_2}$

Since the relation (16) determines φ_{stat}, the optimal value of γ_{stat} may be found, say, from (12):

(17)
$$
\begin{aligned}
\gamma_{\text{stat}} &= \frac{s_b \cos\varphi_{\text{stat}} + c_b \sin\varphi_{\text{stat}}}{s_2 \cos^2\varphi_{\text{stat}} + 2d\cos\varphi_{\text{stat}}\sin\varphi_{\text{stat}} + c_2\sin^2\varphi_{\text{stat}}} \\[2mm]
&= \frac{c_b d^2 - 2s_b c_b d + s_b c_2 + c_b s_2}{d^2 - s_2 c_2}
\end{aligned}
$$

Thus, (16) and (17) solve the problem of determination of φ_{stat} and γ_{stat} by the method of least squares. It is useful to investigate the sensitivity of this problem. Let us consider the following second partial derivatives of $\sigma^2(\gamma, \varphi, t)$:

(18) $\qquad\qquad\qquad a_{11} = \dfrac{\partial^2 \sigma^2(\gamma, \varphi, t)}{\partial\gamma^2}\bigg|_{\gamma=\gamma_{\text{stat}},\ \varphi=\varphi_{\text{stat}}}$

(19) $\qquad\qquad\qquad a_{12} = \dfrac{\partial^2 \sigma^2(\gamma, \varphi, t)}{\partial\gamma\,\partial\varphi}\bigg|_{\gamma=\gamma_{\text{stat}},\ \varphi=\varphi_{\text{stat}}}$

(20) $\qquad\qquad\qquad a_{22} = \dfrac{\partial^2 \sigma^2(\gamma, \varphi, t)}{\partial\varphi^2}\bigg|_{\gamma=\gamma_{\text{stat}},\ \varphi=\varphi_{\text{stat}}}$

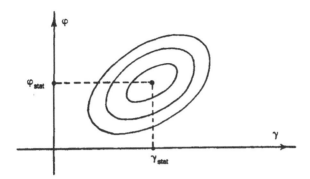

Figure 5.3. Level curves of mean square error $\sigma(\gamma, \varphi, t)$ at fixed t.

Taking into account (12), (13), (16) and (17), it is easy to obtain the following expressions for the second derivatives:

(21)
$$a_{11} = 2(s_2 \cos^2 \varphi_{\text{stat}} + 2d \cos \varphi_{\text{stat}} \sin \varphi_{\text{stat}} + c_2 \sin^2 \varphi_{\text{stat}})$$
$$= \frac{2}{\gamma_{\text{stat}}}(s_b \cos \varphi_{\text{stat}} + c_b \sin \varphi_{\text{stat}})$$

(22) $\quad a_{12} = 2(c_b \cos \varphi_{\text{stat}} - s_b \sin \varphi_{\text{stat}})$

(23) $\quad a_{22} = 2\gamma_{\text{stat}}(s_2 \sin \varphi_{\text{stat}} - 2d \sin \varphi_{\text{stat}} \cos \varphi_{\text{stat}} + c_2 \cos^2 \varphi_{\text{stat}})$

In order to estimate deviation of the mean square error $\sigma(\gamma, \varphi, t)$ under a variation of γ and φ near φ_{stat} and γ_{stat}, expand $\sigma(\gamma, \varphi, t)$ in a neighborhood of the point $\gamma_{\text{stat}}(t)$, $\varphi_{\text{stat}}(t)$:

(24)
$$\sigma(\gamma, \varphi, t) = \sigma_{\text{min}}^2 + a_{11}(t)(\gamma - \gamma_{\text{stat}}(t))^2$$
$$+ 2a_{12}^2(\gamma - \gamma_{\text{stat}}(t))(\varphi - \varphi_{\text{stat}}(t)) + a_{22}(t)(\varphi - \varphi_{\text{stat}}(t))^2$$

in (24) we neglected summands of the third and higher degrees with respect to small differences $\gamma - \gamma_{\text{stat}}$ and $\varphi - \varphi_{\text{stat}}$.

The expansion (24) enables us to estimate sensitivity of the mean square error $\sigma(\gamma, \varphi, t)$ to variations of γ and φ, because the coefficients a_{11}, a_{12} and a_{22} may be found with the help of (21)–(23) from modern coordinates of stars and the coordinates contained in the *Almagest*. It follows from the relation (24) that the level lines of the mean square error are ellipses in the coordinate plane (γ, φ) with the center at $(\gamma_{\text{stat}}, \varphi_{\text{stat}})$ (Figure 5.3). The value of $\sigma(\gamma, \varphi, t)$ at this point is σ_{min}. The directions and the lengths of the axes may be found from standard formulas of analytic geometry; namely, the slope α of one of the axes is determined by

$$(25) \qquad\qquad \tan 2\alpha = \frac{2a_{12}}{a_{11} - a_{22}}$$

and the other axis is perpendicular to the first; the ratio of the axes is equal to the ratio of roots of the quadratic equation

$$\lambda^2 - \lambda(a_{11} + a_{22}) + (a_{11}a_{22} - a_{12}^2) = 0$$

4. Variation of γ_{stat} and φ_{stat} with the a priori date

In the previous section, we fixed the time t; here we will consider variation of γ_{stat} and φ_{stat} with variation of time.

Of course, the behavior of the parameters with variation of t could be obtained from formulas of Section 3, which include $L_i(t)$ and $B_i(t)$ (due to which γ_{stat} and φ_{stat} vary); the variation of $L_i(t)$ and $B_i(t)$ with time has been studied well (see Chapter 1). However, this approach is fairly cumbersome, and in fact is used in our computer calculation of $\gamma_{stat}(t)$ and $\varphi_{stat}(t)$ (see Chapter 6). Here we confine ourselves to a qualitative analysis of behavior of the two functions.

Let us consider the fixed celestial sphere with *fixed* stars in it. Thus, we go back to Ptolemaic views, although for simplicity of reasoning and calculations. The fact is that the stars that have a notable proper motion (several minutes of arc in the 2.5 thousand years we are interested in) are comparatively few, so their existence does not affect much the general picture. In Figure 5.4 we depict the celestial sphere and the true ecliptic in the year t_A (it is useful to compare Figure 5.4 with Figure 5.1). The compiler of the catalog determined the position of the ecliptic inexactly, so the pole P_A of the ecliptic of the catalog is different from the true pole $P(t)$ of the year t. Let $QP(t_A)R$ be the arc of the large circle through $P(t_A)$ and the equinoctial points Q_A and R_A, and let $D(t_A)D'(t_A)$ be the large circle through $P(t)$ perpendicular to $QP(t_A)R$. If we knew the year t, the method of least squares applied as in Section 3 would produce the parameters γ and φ that determine the position of the ecliptic of the catalog in relation to the true ecliptic of the year t. It is obvious from Figure 5.4 that the two angles also determine the position of $P(t_A)$ in relation to P_A, namely, γ is the angle subtended by the arc $P(t)_AP$, and φ is the angle $P_AP(t_A)D'(t_A)$. As we noted in Chapter 1, the position of the ecliptic varies with time, so at a moment t distinct from t_A, the (true) pole is at the point $P(t)$, distinct from $P(t_A)$. The trajectory of the variation of the pole of the

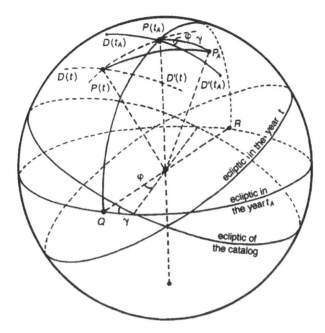

Figure 5.4. Geometry of the angles φ and γ in the celestial sphere.

ecliptic with time is depicted in Figure 5.4 by the dashed line through $P(t)$ and $P(t_A)$. In order to bring the true ecliptic of the year t into coincidence with the ecliptic of the catalog, we need to bring P_A into coincidence with $P(t)$. The length of the arc $P(t)P_A$ is equal to $\gamma_{\text{stat}}(t)$. As for the position of the axis of the turn that generates coincidence, it may be parametrized by the angle $P_A P(t) D'(t)$ where $D(t) D'(t)$ is the arc "parallel" to the arc $D(t) D'(t)$.

In order to see the qualitative behavior of the functions $\gamma_{\text{stat}}(t)$ and $\varphi_{\text{stat}}(t)$, let us consider a "flat" picture where we only depict the process of variation of poles of the ecliptic; this approach is acceptable because the variation surely does not exceed 1°. Transfer the picture near the North pole of the ecliptic from Figure 5.4 to Figure 5.5.

Thus, the true pole of the ecliptic shifts with time, due to oscillation of the ecliptic (see Figure 5.5). Within the interval of time we consider, the variation is about 25′, so we may depict it as a rectilinear segment (the dashed line in Figure 5.5). The motion of the pole of the ecliptic is to a high accuracy uniform, so the distance between $P(t)$ and $P(t_A)$ may be assumed to be equal to $v \cdot (t - t_A)$ where v is the velocity of motion of the ecliptic (approximately, 1′ per century). Recall that P_A is the point where the compiler of the catalog placed the pole because of the error in determination of the ecliptic; if the

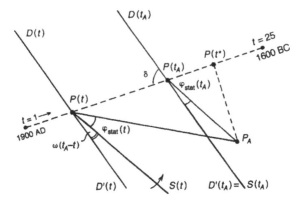

Figure 5.5. Trajectory of the true motion of the ecliptic in the celestial sphere due to pre-cession. Geometrical sense of the parameters γ and φ dependence on the a priori date. The point P_A depicts the position of the pole given in the *Almagest*.

perpendicular dropped from P_A on the dashed line depicting the motion of the pole of the ecliptic meets this line at some $t^* > t_A$ (as shown in Figure 5.5), then the compiler's error "renders the catalog older", because the ecliptic of the year t^* is closer to the ecliptic of the catalog; in the opposite case (if the perpendicular meets the dashed line at $t^* < t_A$), the error "rejuvenates" the catalog. To give some notion about the real values of the quantities in question, the distance between the pole of the ecliptic of 1900 AD, $P(0)$ and of 1 AD, $P(19)$ is about 20'; $\gamma_{\text{stat}}(t)$ for the *Almagest* is also approximately this value.

As we know, $\gamma_{\text{stat}}(t_A)$ is equal to the length of the arc $\overline{P(t_A)P_A}$, and $\varphi_{\text{stat}}(t_A)$ to the angle $P_A P(t_A) D'(t_A)$. Similarly,

$$(1) \qquad\qquad \gamma_{\text{stat}}(t) = \overline{P(t)P_A}$$

where the bar denotes the (angular) length of the arc. But the angle $P_A P(t) D'(t)$ is not equal to $\varphi_{\text{stat}}(t)$, because by the moment t the equinoctial axis has displaced by $\omega \cdot (t - t_A)$ where ω is the angular velocity of precession (about 50" per year; see Chapter 1). This displacement corresponds to the angle $D'(t)P(t)S(t)$ in Figure 5.5. Thus,

$$(2) \qquad\qquad \varphi_{\text{stat}}(t) = \angle P_A P(t) S(t)$$

($\angle$ stands for "angle"); we have also

$$(3) \qquad\qquad \angle D'(t)P(t)S(t) = \omega \cdot (t - t_A)$$

In order to avoid cumbersome notation, we put

(4) $x(t) = \overline{P(t)P(t_A)}$

(5) $y = P(t_A)P(t^*)$

(6) $\psi(t) = \angle P_A P(t) D'(t)$

(7) $z = P_A P(t^*)$

(8) $\delta = \angle D(t_A)P(t_A)P(t)$

The quantity $\gamma_{\text{stat}}(t)$ may be called *the error in determination of the ecliptic* (in the *Almagest* it is about 20'). The angle δ is constant and is equal to the angle the direction of motion of the pole of the ecliptic makes with the line $D(t_A)D'(t_A)$. Clearly,

(9) $z = \gamma_{\text{stat}}(t_A)\sin(\delta - \varphi_{\text{stat}}(t_A))$

(10) $\gamma = \gamma_{\text{stat}}(t_A)\cos(\delta - \varphi_{\text{stat}}(t_A))$

Since $x(t) = v \cdot (t - t_A)$, it readily follows from Figure 5.5 that

(11) $\gamma_{\text{stat}} = \left[(v\cdot(t-t_A)+\gamma)^2+z^2\right]^{1/2} = \left[\gamma_{\text{stat}}^2(t_A)+2\gamma v\cdot(t-t_A)+v^2(t-t_A)^2\right]^{1/2}$

Clearly, the function in the right side of (11) attains its minimum value at $t = t^*$; if we consider the case when $|t - t_A| \ll |t_A - t^*|$, the function $\gamma_{\text{stat}}(t)$ is practically linear:

(12) $\gamma_{\text{stat}}(t) \approx \gamma_{\text{stat}}(t_A) + v\cos(\delta - \varphi_{\text{stat}}(t_A))(t_a - t)$

The function $\varphi_{\text{stat}}(t)$ is also easy to determine:

(13) $\varphi_{\text{stat}}(t) = \delta + \omega \cdot (t_a - t) - \arctan\left(\dfrac{z}{y + v \cdot (t_a - t)}\right)$

Again, if $|t_A - t| \ll |t_A - t^*|$, then we may use the linear approximation

(14) $\varphi_{\text{stat}}(t) \approx \varphi_{\text{stat}}(t_A) + \left[\omega + \dfrac{v\sin(\delta - \varphi_{\text{stat}}(t_A))}{\gamma_{\text{stat}}(t_A)}\right](t_A - t)$

Of course, the above formulas only give a general view of the behavior of $\gamma_{\text{stat}}(t)$ and $\varphi_{\text{stat}}(t)$. Figure 5.6 displays the functions obtained from (11) and (13). Naturally, the particular form of the functions depends upon the errors made by the compiler of the catalog, that is, on $\gamma_{\text{stat}}(t_A)$ and $\varphi_{\text{stat}}(t)$.

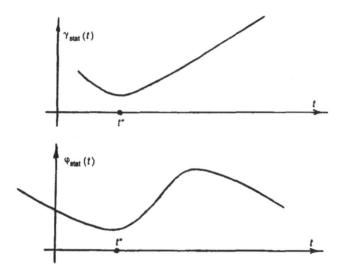

Figure 5.6. Approximate form of the functions $\gamma_{\text{stat}}(t)$ and $\varphi_{\text{stat}}(t)$.

The relations (11) and (13) also determine the form of the function $\beta_{\text{stat}}(t)$; see (2.1).

Let us now describe the geometrical sense of the above constructions. If we consider coordinates of a group of stars at the moment t, then compensation for the group error $\varphi_{\text{stat}}(t)$, $\gamma_{\text{stat}}(t)$ (that is, the turn of the group through the angle $\gamma(t)$ about the axis that makes the angle $\varphi_{\text{stat}}(t)$ with the equinoctial axis), brings the pole of the catalog P_A into coincidence with the true pole $P(t)$. Of course, this does not eliminate latitudinal deviations, because the catalog also contains random errors; their *mean* is zero, so *they do not displace the pole of the ecliptic* (to be precise, displace by a small distance; the more stars in the group considered, the less is the displacement).

The displacement of P to the point $P(t)$ is *uniquely* decomposed into the displacements $P_A \rightarrow P(t_A)$ and $P(t_A) \rightarrow P(t)$. The parameters $\gamma_{\text{stat}}(t_A)$ and $\varphi_{\text{stat}}(t_A)$ that parametrize the first of the two displacements may be treated as the error of the observer, that is, the error the compiler of the catalog had made in determination of the ecliptic. The second displacement is due to oscillation of the ecliptic plane; it may be computed from Newcomb's theory.

It follows also that if we denote by $\Delta B_i(t)$ the latitudinal deviation of the ith star computed for the moment of time t, and by $\Delta B_i^0(t) = \Delta B_i(t) - \gamma_{\text{stat}}(t)\sin(L_i(t) + \varphi_{\text{stat}}(t))$ the latitudinal deviation of the star after compensation for the systematic error, then *for the collection of absolutely fixed stars, the deviations $\Delta B_i(t)$ do not vary with t and are due to the random errors made by the compiler of the catalog in measurement of latitudes.* This is not the case if the collection includes *moving* stars; their deviations $\Delta B_i(t)$ depend upon t.

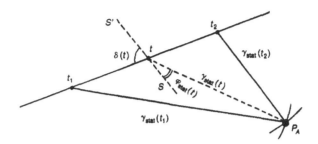

Figure 5.7. Finding the values of $\gamma_{\text{stat}}(t)$ and $\varphi_{\text{stat}}(t)$.

The form of these dependencies is determined by individual random errors, as well as by directions of proper motion. In particular, in the year t_A (unknown to us), $\Delta B_i(t)$ is equal to the individual random error of the star i. It is natural to expect that if the star moves fast and is well-measured, then $|\Delta B_i(t)|$ attains its minimum value in a neighborhood of t_A; the length of the neighborhood depends on the velocity and the direction of the proper motion, and even for the fastest stars (say, Arcturus) amounts to hundreds of years.

An important conclusion is that *determination of the pole of the ecliptic of the catalog P_A only requires the values of γ_{stat} at two moments of time t_1 and t_2*. Indeed, from the theory of Newcomb (see Chapter 1) it is easy to find the velocity v of displacement of the pole of the ecliptic. Fix two arbitrary distinct moments of time t_1 and t_2 (see Figure 5.7) and find the values $\gamma_{\text{stat}}(t_1)$ and $\gamma_{\text{stat}}(t)$ from (3.16); draw the line along which the pole of the ecliptic moves and mark the points t_1 and t_2. Choose the scale so that the distance between the marked points is equal to $v \cdot (t_2 - t_1)$. The position of the pole of the ecliptic P_A may be found as the intersection of the two circumferences with the centers at the points t_i and the radii $\gamma_{\text{stat}}(t_i)$, $i = 1, 2$. It is now obvious from Figure 5.7 how to determine the values $\gamma_{\text{stat}}(t)$ and $\varphi_{\text{stat}}(t)$ for any t. We should note only that the line SS' from which the angle $\varphi_{\text{stat}}(t)$ is counted meets the trajectory of the pole of the ecliptic at an angle $\delta(t)$, also determined from Newcomb's theory. The astronomic sense of the line SS' is quite obvious; it is the "rectilined" segment of the large circle of the celestial sphere that passes through the pole of the ecliptic in the year t, $P(t)$ and is perpendicular to the large circle through $P(t)$ and the spring equinoctial point of the year.

Similarly, the values of $\varphi_{\text{stat}}(t)$ at two distinct values of t determine the values of $\gamma_{\text{stat}}(t)$ and $\varphi_{\text{stat}}(t)$ for *all* t.

However, we will work with the angle γ, for it has an obvious sense, the error in the inclination of the ecliptic to the equator. Note that this inclination is fixed, say, in an armillary sphere (see Chapter 1), for which, consequently,

γ is an instrumental error. Furthermore, the choice of γ will be justified below from the statistical point of view.

5. Statistical properties of the estimates γ_{stat} and φ_{stat}

In this section we approach the problem of estimating parameters γ and φ that determine the systematic error of the catalog as a statistical one. To that end we assume the following. Suppose that the compiler of the catalog for the year t_A had made the systematic error determined by the parameters γ_{stat} and φ_{stat}. Suppose that, moreover, the latitude of each coordinate was affected by a random perturbation ξ_i (the individual observational error) with zero mean, that is, $\mathbf{E}\,\xi_i = 0$. We assume that the random errors ξ_i corresponding to various stars have the same distribution. Let $\sigma^2 = \mathbf{E}\,\xi^2$ be the variance of the random variable ξ^2 (strictly speaking, we do not know the value of σ^2). Then the latitude given in the catalog may be written in the form

(1) $$b_i = B_i(t_A) - \gamma_A \sin(L_i(t_A) + \varphi) + \xi_i$$

Thus, from the statistical point of view the catalog is a sampling of size N of realizations of the random variables $\{b\}_{i=1}^N$ of the form (1), use to which we must find the statistical estimates γ_A and φ_A, respectively, of the parameters γ and φ, and also to estimate the variance σ (the mean square error of the observations). We will restrict the problem and will study the statistical properties of the estimates $\varphi = \varphi_{\text{stat}}$ and $\gamma = \gamma_{\text{stat}}$ produced by the least squares method and given by (3.16) and (3.17); the main attention will be given to the estimate of γ_A, for the reasons explained in the end of the next section. Let us reduce (1) to the form traditional for regression analysis. The observational error $\Delta b_i = B(t_A) - b_i$ is a random variable whose mean is equal to $\gamma_A \sin(L_i(t) + \varphi_A)$ and whose variance is σ^2. The curve $Y(x) = \gamma_A \sin(x + \varphi_A)$ is usually called the *regression curve*.

By our assumption, the deviations Δb_i are random, so the estimates γ_{stat} and φ_{stat} that ensue from (3.16) and (3.17) are also random variables. Let us investigate their statistical properties and their relation with the true (unknown) values φ_A and γ_A.

Let us write φ_{stat} and γ_{stat} in a special form; namely, substitute $\gamma_A \sin(L_i(t) + \varphi_A) - \xi_i$ for Δb_i in the expressions for s_b and c_b; now (3.16) and (3.17) assume the form

(2) $$\tan \varphi_{\text{stat}} = \frac{\tan \varphi_A + \left[(R/N) \sum_{i=1}^N \xi_i (s_2 \cos L_i(t_A) - d \sin L_i(t_A))\right]}{1 + \left[(R/N) \sum_{i=1}^N \xi_i (c_2 \sin L_i(t_A) - d \cos L_i(t_A))\right]}$$

(3) $$\gamma_{\text{stat}} = \gamma_A - \frac{(1/N) \sum_{i=1}^N \xi_i (\sin L_i(t_A) + \tan \varphi_A \cos L_i(t_A))}{(s_2 + 2d \tan \varphi_A + c_2 \tan^2 \varphi_A) \cos \varphi_A}$$

where

(4)
$$R = \left(\gamma_A(d^2 - s_2 c_2)\cos\varphi_A\right)^{-1}$$

From $\mathbf{E}\,\xi_i = 0$ follows

(5)
$$\mathbf{E}\,\gamma_{\text{stat}} = \gamma_A$$

whence the estimate for γ_{stat} is unbiased. As for the variance $\mathbf{D}\,\gamma$ of the estimate γ_{stat}, it is given by

(6)
$$\mathbf{D}\,\gamma = \frac{\sigma^2}{N(s_2\cos^2\varphi_A + 2d\cos\varphi_A\sin\varphi_A + c_2\sin^2\varphi_A)}$$

If the individual errors ξ_i are normally distributed, then so is the variable γ_{stat}, and the first two moments (5) and (6) completely determine its characteristics. In the sequel, we will use this fact for constructing confidence interval for the value of γ_A.

The analysis of the estimate φ_{stat} is a bit more complicated. We will use the following relation, easily derivable from (2):

(7)
$$\tan\varphi_{\text{stat}} - \tan\varphi_A$$
$$= \frac{(R/N)\sum_{i=1}^{N}\xi_i\big((s_2 + d\tan\varphi_A)\cos L_i(t_A) - (d + c_2)\sin L_i(t_A)\big)}{1 + (R/N)\sum_{i=1}^{N}\xi_i\big(c_2\sin L_i(t_A) - d\cos L_i(t_A)\big)}$$

and the fact that at large N the second summand in the denominator of the right side of (7) is small. Indeed, this quantity is a random variable with zero mean and variance equal to $\dfrac{\sigma^2 c_2}{N\gamma^2(s_2 c_2 - d^2)\cos^2\varphi_A}$. If, moreover, ξ_i are normally distributed, then so is the second summand in the denominator of the right side of (7). It follows that for the catalog of the *Almagest*, already at $N = 30$ the probability that the denominator in the right side of (7) is negative does not exceed $5\cdot 10^{-3}$; with the growth of N this probability rapidly decreases: $P_{50} \le 2.5\cdot 10^{-4}$, $P_{80} \le 4\cdot 10^{-6}$, $P_{100} \le 3\cdot 10^{-7}$, $P_{200} \le 8\cdot 10^{-13}$, $P_{300} \le 2.5\cdot 10^{-18}$. It follows from (7) that generally, $\mathbf{E}\tan\varphi_{\text{stat}} \neq \tan\varphi_A$, but it is easy to derive the distribution function $F(x)$ of the random variable $\tan\varphi_{\text{stat}} - \tan\varphi_A$ (necessary for finding the confidence interval for φ_A), if we neglect the case that the denominator of (7) is negative, the probability of which is low:

(8)
$$F(x) = \mathbf{P}(|\tan\varphi_{\text{stat}} - \tan\varphi_A| < x) = \mathbf{P}(\eta_x < x)$$

where the random variable η_x is of the form

(9)
$$\eta_x = \frac{R}{N} \sum_{i=1}^{N} \xi_i \big((s_2 + d(\tan \varphi_A + x) \cos L_i(t_A))$$

$$- (d + (c_2 \tan \varphi_A + x)) \sin L_i(t_A) \big)$$

Consequently, if the variables ξ_i are normally distributed, then so is η_x, with the mean 0 and the variance

(10)
$$\mathbf{D}\,\eta_x = \frac{R^2 \sigma^2}{N} (c_2 s_2 - d^2)(s_2 + 2d(x + \tan \varphi_A)^2)$$

Hence,

(11)
$$F(x) = \Phi(x/\sqrt{\mathbf{D}\,\eta_x})$$

where

(12)
$$\Phi(x) = \frac{1}{\sqrt{2\pi}} \int_{-\infty}^{x} \exp\left(-\frac{1}{2}u^2\right) du$$

The above variables φ_{stat} and γ_{stat} are the so-called *point estimates* for the unknown parameters φ_A and γ_A.

Since we have found the distributions of these variables, we now can investigate the accuracy of the estimates found. We will estimate the accuracy in terms of *confidence intervals*.

In mathematical statistics, the problem of finding confidence intervals comes from the situation which we will explain in terms of the example of the estimate of γ_A. The γ_A is a *nonrandom* (*determined*) quantity, the error the compiler made as he created the catalog. Applying the least squares method, we obtain a *random* variable γ_{stat}. Of course, we would like to know, what bounds can we find for γ_A given a realization of γ_{stat}. In order that the bounds be nontrivial, we have to set the *confidence level* $1 - \varepsilon$ where ε is the probability of an error that we admit. Since the variable γ is normally distributed with the parameters as in (5) and (6), we have for all $x > 0$,

(13)
$$P(|\gamma_{\text{stat}} - \gamma_A| < x) = \Phi(\sqrt{\mathbf{D}\gamma} \cdot x) - \Phi(-\sqrt{\mathbf{D}\gamma} \cdot x)$$

Find x_ε from the equation

(14)
$$\Phi(\sqrt{\mathbf{D}\gamma} \cdot x_\varepsilon) - \Phi(-\sqrt{\mathbf{D}\gamma} \cdot x_\varepsilon) = 1 - \varepsilon$$

or, equivalently, from the equation

(15) $$\Phi(-\sqrt{\mathbf{D}\gamma} \cdot x_\varepsilon) = \varepsilon/2$$

Then the interval

(16) $$I_\gamma(\varepsilon) = (\gamma_{\text{stat}} - x_\varepsilon, \gamma_{\text{stat}} + x_\varepsilon)$$

is the *confidence interval* for γ_A, because

(17) $$\mathbf{P}(|\gamma_{\text{stat}} - \gamma_A| \geq x_\varepsilon) = \varepsilon$$

To find x_ε, we need to use the value of $\mathbf{D}\gamma$, which depends on the unknown parameters σ^2 and φ_A. As is usually done in mathematical statistics, we substitute for σ^2 in (6) the *residual variance* (see (3.8)):

(18) $$\sigma^2(\gamma_{\text{stat}}, \varphi_{\text{stat}}, t_A) = \frac{1}{N} \sum_{i=1}^{N} (\Delta B_i(t_A) - \gamma_{\text{stat}} \sin(L_i(t_A) + \varphi_{\text{stat}}))^2$$

and for φ_A its approximation $\varphi_{\text{stat}}(t_A)$. Since the moment of compilation of the catalog t_A is also unknown, we need to do all this for all moments of time t, in order to estimate the *conditional systematic error* $\gamma_{\text{stat}}(t)$, $\varphi_{\text{stat}}(t)$, provided that the true date of compilation is t.

A similar procedure may be applied to finding the confidence interval for φ_A with the confidence level ε; the resulting interval is

(19) $$I_\varphi(\varepsilon) = \left[\varphi_{\text{stat}} - \frac{y_\varepsilon}{1 + \tan^2 \varphi_{\text{stat}} - y_\varepsilon \tan \varphi_{\text{stat}}} \right.,$$
$$\left. \varphi_{\text{stat}} + \frac{y_\varepsilon}{1 + \tan^2 \varphi_{\text{stat}} + y_\varepsilon \tan \varphi_{\text{stat}}} \right]$$

where y_ε is the root of the equation

(20) $$F(y_\varepsilon) - F(-y_\varepsilon) = 1 - \varepsilon$$

and the distribution function F is as in (11).

6. Conclusions

1.   The group error for a star configuration reduces to a displacement of the configuration as a whole in the celestial sphere. If we are only interested

in latitudinal deviations, the displacement may be determined in terms of two parameters, γ and φ (or γ and β).

2. The latitudinal deviations of the catalog may be diminished by compensation for the group errors.

3. If the group errors coincide in a large part of the catalog, then the common error is called *systematic*, and may be found by application of a statistical approach.

Under the condition that the catalog has been compiled in the year t, the values of $\varphi(t)$ and $\gamma(t)$ may be estimated by the method of least squares; the resulting estimates $\varphi_{stat}(t)$ and $\gamma_{stat}(t)$ are as in (3.16) and (3.17).

4. The functions $\gamma(t)$ and $\varphi(t)$ may be restored from the values of $\gamma(t)$ at two distinct values of t.

5. Under the assumption that random measurement errors are normally distributed, the confidence intervals for the true values of $\varphi(t)$ and $\gamma(t)$, $I_\varphi(t)$ and $I_\gamma(t)$, are given by (5.19) and (5.16).

Chapter 6

Statistical Properties and the Accuracy of the Catalog of the *Almagest*

1. Introductory remarks

We already know that the analysis of errors in the catalog and the related problem of the accuracy of measurements are crucial for dating. The failures of the attempts reviewed in Chapter 3 may be explained by the fact that they did not involve any analysis of accuracy, both of the entire catalog and of measurements of separate groups of stars. A preliminary rough analysis of the accuracy we have carried out in Chapters 2 and 4. In Chapter 2, on the basis of a generalization of results obtained by various researchers of the *Almagest*, we have revealed the division of the star atlas into seven domains that differ in the rate of doubtfully identifiable stars. In Chapter 4 we also considered the problem of identification of fast stars in a modern catalog with stars in the *Almagest*.

The main tools for what we are going to do in this chapter are the methods exposed in Chapter 5. Using them, we first prove that the seven domains really have different accuracy characteristics. Namely, they differ both in systematic and random errors. For these domains, we will figure out the errors in the determination of poles of the ecliptic and the mean square errors of measurements. Further, we find the confidence intervals for the estimates γ_{stat} and φ_{stat} of the parameters of systematic errors. Then we analyze comparatively small areas of the sky, constellations and vicinities of particular stars. The aim of this analysis is to check that the previously found values of γ_{stat} and φ_{stat} are indeed systematic errors of the catalog, but not the result of superposition of several group errors characteristic for various groups of stars.

As a result, we educe the well-measured domain of the sky, in fact, fairly large, which we will further use for dating the catalog.

2. Seven domains of the starry sky

1.   In Chapter 2 we have described seven domains in the sky, see Figure 6.1. We have considered 864 stars in these domains; for the reasons explained in Chapter 2, we excluded from consideration the stars listed in the informatas, as well as outlies and doubtfully identified stars. Table 6.1 indicates what particular stars are attributed to what particular domains. Namely, each line of the table, corresponding to a domain exhibits the Baily numbers of the stars attributed to the domain.

We have also marked in Figure 6.1 the stars that have proper names in the *Almagest*. It is obvious that the domain A is clearly delineated by the named stars. This apparently indicates that the compiler attributed a special significance to the domain (see Chapter 2). The domain A is of special importance for us; it contains the pole of the world N and the pole of the ecliptic P. If the named stars that border the domain served as the reference stars, then Ptolemy proceeded inwards in the domain A, starting from these stars and measuring the coordinates of the rest of the stars in reference to the marked named stars. In this process errors of measurement could accumulate, so it is natural to expect that the stars in the domain A outside the zodiac are measured on average a little worse than the zodiacal stars. Six named stars of the *Almagest* lie either in the zodiac or immediately near it (Procyon).

2.   Let us begin with finding the positions of the pole of the ecliptic appropriate to each of the seven domains. As shown in Chapter 5, the positions are determined by the parameters γ_{stat} and φ_{stat}, which can be found from the method of least squares. So, let us calculate from (5.3.16) and (5.3.17) the values of $\gamma_{stat}(t)$ and $\varphi_{stat}(t)$ for each of the seven domains and find the corresponding positions of the pole of the ecliptic; see Figure 6.2, where the true positions $P(t)$ of the pole of the ecliptic are depicted for the values of t ranging from 1 to 25. The length of the segment that connects the position of the pole of the ecliptic appropriate, say, for the domain B with, say, $P(10)$ is equal to $\gamma_{stat}^B(10)$, and the angle counted from the line denoted by $D'(10)$ (cf. Figures 5.4 and 5.5) is equal to $\varphi_{stat}^B(10)$; of course, similar magnitudes for other domains and other values of t have similar sense.

In Table 6.2 we give exact values of $\gamma_{stat}(18)$ and $\varphi_{stat}(18)$ for each of the domains. These values fix the position of the pole of the ecliptic appropriate to the domain (as well as any pair of values $\gamma_{stat}(t)$ and $\varphi_{stat}(t)$; see Section 5.4). Besides, Table 6.2 contains the initial mean square deviations $\sigma_{init}(18)$

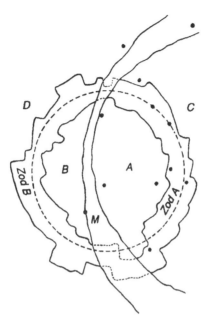

Figure 6.1. Seven domains of the starry sky of the *Almagest*.

and the residual mean square deviations σ_{min} that ensue after compensation for the systematic errors (see (5.3.8) and (5.3.9)). We have shown in Section 5.4 that σ_{min} does not depend upon the moment of time t (if the stars in question are fixed), but is determined by the position of the "observer's pole of the ecliptic" of the group of stars. The value $t = 18$ in Table 6.2 is chosen because it is the traditional date for the *Almagest*. Further, in Table 6.2 we give $p_{init}(18)$, the rate of the stars that at $t = 18$ had latitudinal deviations below 10′, and p_{min}, the rate of the stars that have latitudinal deviation below 10′ after compensation for the systematic error.

The positions of poles in Figure 6.2 show that the systematic error in each group except C renders the catalog older, even in comparison with the time of Hipparchus. The minimum of systematic error for the domain C is attained at $t \approx 10$ ($\approx$ 900 AD). However, as we already noted, *the position of the "observer's pole of the ecliptic" has nothing to do with the date of compilation of the catalog. The position only indicates the character and magnitude of the systematic error made by the observer.* It also follows from Figure 6.2 that the positions of the pole appropriate to the domains A, *Zod A* and *Zod B* are rather near, which may mean that the three groups carry the same group error; we will consider this question in details below, in our analysis of individual

Table 6.1.

Domain of the sky	Baily's numbers (before exclusion)	Number of stars in the domain left after exclusion
A	1–158, 424–569	249
B	286–423, 570–711	262
C	847–977	116
D	712–846, 998–1028	143
M	159–285	94
Zod A	424–569	124
Zod B	362–423, 570–711	168

constellations. The position of the pole of the ecliptic appropriate to the domain B is also not far from the three positions. The positions appropriate to the domains D and M are a little further; apparently, the systematic error in these domains is different from, say, the one in the domain A. The domain C seems to be an outlie.

3.   In order to make the above argument exact, we will use the notion of confidence intervals; see Section 5.5. We find the functions $\gamma_{\text{stat}}(t)$ and $\varphi_{\text{stat}}(t)$, $1 \le t \le 25$, for each of the seven domains, and display in their graphs the confidence intervals $I_\gamma(\varepsilon)$ and $I_\varphi(\varepsilon)$ for $\varepsilon = 0.1$ (see (5.5.16) and (5.5.19)). The graphs are shown in Figures 6.3–6.9. More complete information about the lengths of confidence intervals for various values of ε and the values $t = 7$ and $t = 18$ is given in Table 6.3. Recall (see Section 5.5) that the point $\gamma(t)$ is the center of the interval $I_\gamma(\varepsilon)$. The interval $I_\varphi(\varepsilon)$ need not be symmetric with center at the point $\varphi_{\text{stat}}(t)$, but this asymmetry is negligible (for our purposes), so we will think that $\varphi_{\text{stat}}(t)$ is in the middle of the confidence interval. In Table 6.3 x_ε^γ denotes the radius of $I_\gamma(\varepsilon)$, and x_ε^φ the radius of $I_\varphi(\varepsilon)$.

The data in Table 6.3 lead to the following conclusions. The stars in *Zod A* are measured best; namely, compensation for the systematic error reduces the mean square error in this group down to $12'.8$, rendering latitudinal deviations of about 64% stars below 10'. Recall that seven named stars are contained in this domain or immediately near it (Aquila, Regulus, Arcturus, Procyon, Vindemiatrix, Spica, and Antares).

The next in accuracy is the group A, where compensation for the group error lowers the mean square latitudinal deviation to $16'.5$, and increases the rate of stars with latitudinal deviations below 10' to more than 50%.

The confidence intervals $I_\varphi(\varepsilon)$ and $I_\gamma(\varepsilon)$ for the domains *Zod A* and *A* have similar lengths (see Table 6.3), although the accuracy in *Zod A* is higher than in A. This is due to the difference in the number of stars—the less is the number of stars, the greater is the length of the confidence interval (and the higher is the accuracy, the less is the length of the confidence interval).

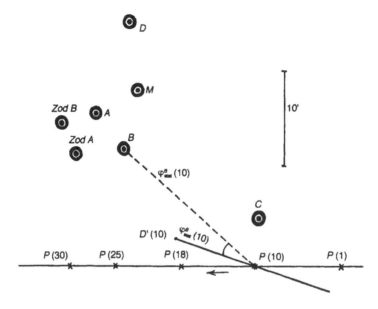

Figure 6.2. Relative positions of the moving true pole of the ecliptic and the poles of the ecliptic appropriate to each of the seven domains in the sky of the *Almagest*.

Table 6.2.

	A	B	C	D	M	Zod A	Zod B
$\gamma_{stat}(18)$	18.5	13.6	9.7	26.6	19.4	16.4	20.0
$\varphi_{stat}(18)$	34.0	−34.5	−122.5	−52.7	−50.5	−21.7	−23.5
$\sigma_{init}(18)$	20.5	21.8	23.4	27.3	23.0	17.7	24.0
σ_{min}	16.5	19.2	22.5	24.4	20.5	12.8	19.3
$P_{init}(18)$	36.5%	35.5%	33.6%	28.7%	37.2%	30.6%	30.9%
P_{min}	50.6%	43.5%	43.1%	35.7%	45.7%	63.7%	44.0%

The data in Table 6.3 confirm the claimed accuracy 10′ (at least, in latitudes).
    The next in accuracy groups of stars are *B* and *Zod B*. Their characteristics are quite similar: mean residual square error is about 19′, and the stars with latitudinal deviations below 10′ constitute about 44%. Although positions of the pole of the ecliptic appropriate to these groups are quite close to the positions appropriate to *A* and *Zod A*, they lie in the confidence intervals for the latter only at as small values of ε as about 0.01; this means that the systematic errors in *B* and *Zod B* apparently differ from the ones in *A* and *Zod A*. Furthermore, the accuracy of measurement of stars in *A* and *Zod A* is

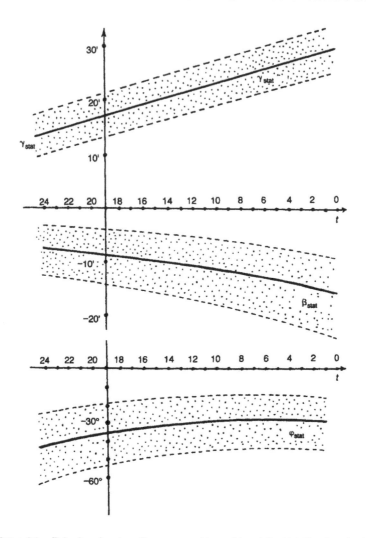

Figure 6.3. Behavior of systematic errors $\gamma_{stat}(t)$, $\varphi_{stat}(t)$ and $\beta_{stat}(t)$ in the domain A.

much better than that in B and $Zod\ B$. Below, we will give more arguments for this statement.

The stars in the domains C, D and M are measured worse than in the domains A and B. Furthermore, the values of γ_{stat} and φ_{stat} for these domains lie in the confidence intervals of similar parameters for A, $Zod\ A$, B and $Zod\ B$ only at rather small values of ε. This means that we are to admit the values

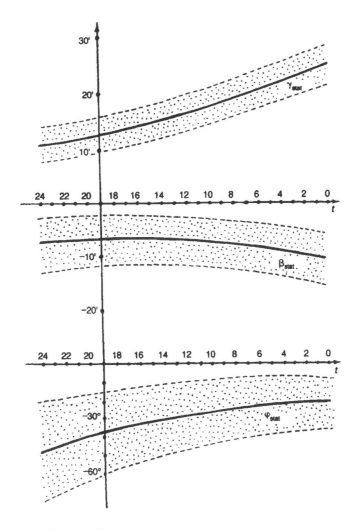

Figure 6.4. Behavior of systematic errors $\gamma_{stat}(t)$, $\varphi_{stat}(t)$ and $\beta_{stat}(t)$ in the domain B.

of systematic errors in these domains being different from the ones in A, $Zod\ A$, B and $Zod\ B$.

As we analyzed Tables 6.2 and 6.3, we already encountered the question: Which values of the mean square error are to be treated as large, and which as small?

In order to answer this question, we will use the analysis of sensitivity carried out in Chapter 5. Our further argument is illustrated in Figure 6.10.

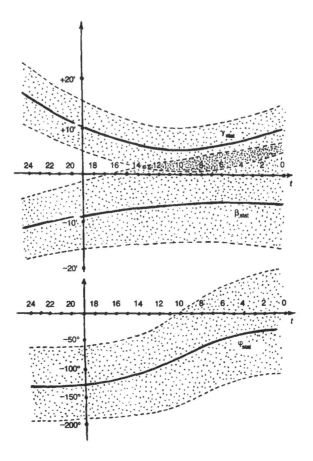

Figure 6.5. Behavior of systematic errors $\gamma_{stat}(t)$, $\varphi_{stat}(t)$ and $\beta_{stat}(t)$ in the domain C.

Let us draw the ellipsoidal level curves of the function $\sigma^2(\gamma, \varphi, 3t)$ in the plane with coordinates (γ, φ) (see (5.3.24)). Draw in the same plane the rectangle with the sides $I_\gamma(\varepsilon)$ and $I_\varphi(\varepsilon)$ (the dotted rectangle in Figure 6.10).

The probability that the true systematic error (γ, φ) lies in the rectangle is not less than $1 - 2\varepsilon$. Let us find $\sigma^2_{max}(\varepsilon) = \max \sigma^2(\gamma, \varphi, t)$ where the maximum is taken over all values of (γ, φ) in $R(\varepsilon)$. The ensuing value of $\sigma^2_{max}(\varepsilon)$ determines the admissible (with the confidence level $1 - 2\varepsilon$) mean square latitudinal deviation, and the difference $\sigma_{max}(\varepsilon) - \sigma_{min}$ the admissible increment of the mean square deviation due to inexactness of the parameters γ and φ that determine the systematic error.

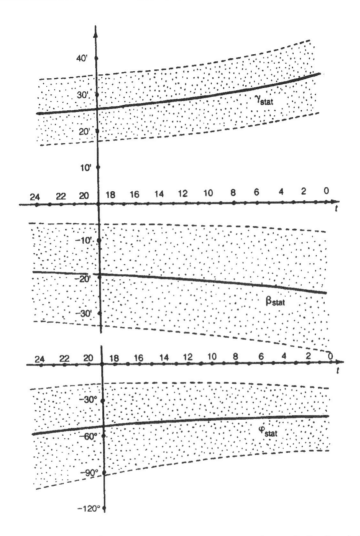

Figure 6.6. Behavior of systematic errors $\gamma_{stat}(t)$, $\varphi_{stat}(t)$ and $\beta_{stat}(t)$ in the domain D.

In Table 6.4 we give the values of a_{11}, a_{12} and a_{22} that determine the level lines of mean square errors (see (5.3.24), where γ is to be expressed in minutes of arc, and φ in degrees of arc) for the domains A and $Zod\ A$ at $t = 18$, and the values of $\Delta\sigma = \sigma_{max}(\varepsilon) - \sigma_{min}$ calculated for the "extreme" values $\varepsilon = 0.1$ and $\varepsilon = 0.005$. Note that the *values thus obtained vary but slightly with time;* they demonstrate an obvious distinction in accuracy of the domains A and $Zod\ A$ from the domains B and $Zod\ B$. Indeed, even at $\varepsilon = 0.01$ *the mean square*

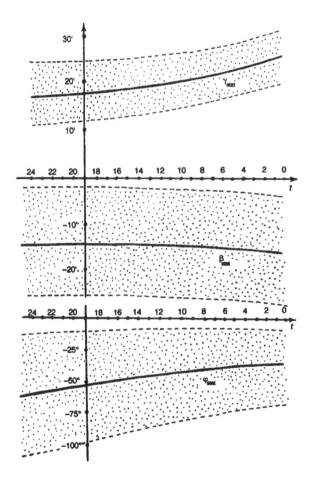

Figure 6.7. Behavior of systematic errors $\gamma_{\text{stat}}(t)$, $\varphi_{\text{stat}}(t)$ and $\beta_{\text{stat}}(t)$ in the domain M.

error in the confidence domain for Zod A is never as large as the minimum error in the domains B and Zod B. A similar statement is true for the domain A: although $\sigma^A_{\max}$ may be greater than $\sigma^B_{\min}$, this only occurs at $\varepsilon \le 0.01$. At the rest of the values of ε the errors in A and B are to be treated as essentially different (distinguishable by statistical criteria). Note that similarly, the stars in Zod A have the accuracy clearly distinct from the stars in A: for *all* values of ε, $\sigma_{\max}$ for Zod A is less than $\sigma_{\min}$ for A.

Further, Table 6.3 shows that the determination of φ_{stat} is not sufficiently stable, especially for the "bad" domains C, D and M, which can be seen

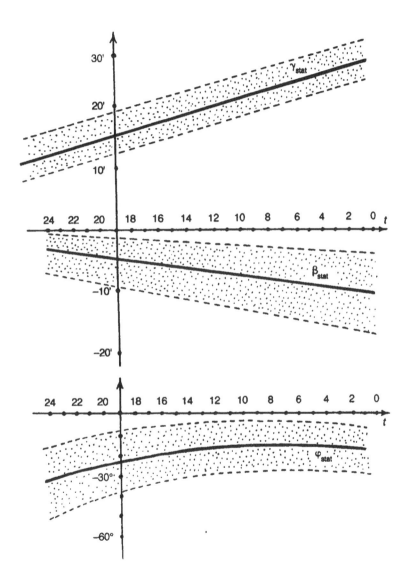

Figure 6.8. Behavior of systematic errors $\gamma_{stat}(t)$, $\varphi_{stat}(t)$ and $\beta_{stat}(t)$ in the domain Zod A.

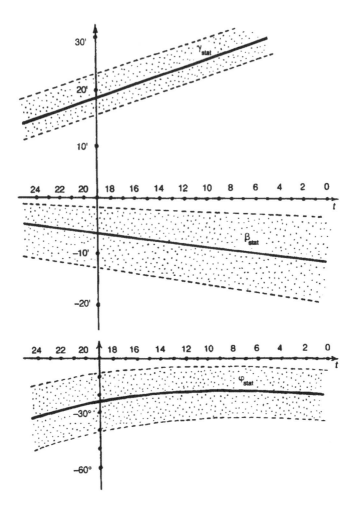

Figure 6.9. Behavior of systematic errors $\gamma_{\text{stat}}(t)$, $\varphi_{\text{stat}}(t)$ and $\beta_{\text{stat}}(t)$ in the domain *Zod B*.

from the lengths of confidence intervals $I_\varphi(\varepsilon)$. For example, the length of the interval for the domain C is greater than $180°$!

3. Analysis of individual constellations

1. In the previous section, we have found, with the help of methods of mathematical statistics, the poles of the ecliptic appropriate to large collec-

Table 6.3.

Domain	$t = 7$				$t = 18$			
	$\varepsilon = 0.1$	$\varepsilon = 0.05$	$\varepsilon = 0.01$	$\varepsilon = 0.005$	$\varepsilon = 0.1$	$\varepsilon = 0.5$	$\varepsilon = 0.01$	$\varepsilon = 0.005$
A	$x_\varepsilon^\gamma = 2.6$	3.1	4.1	4.5	2.7	3.2	4.2	4.6
	$x_\varepsilon^\varphi = 11.7$	14.0	18.3	20.0	16.6	19.8	25.9	28.4
B	2.7	3.2	4.2	4.6	2.6	3.1	4.0	4.4
	14.7	17.4	22.8	25.0	22.1	26.2	34.4	37.6
C	4.6	5.5	7.2	7.9	5.1	6.0	7.9	8.7
	91.1	108.2	141.9	155.2	60.7	72.2	94.7	103.5
D	6.3	7.4	9.8	10.7	7.2	8.6	11.3	12.3
	28.3	33.6	44.1	48.2	37.8	44.9	58.9	64.4
M	5.4	6.4	8.5	9.2	6.5	7.7	10.1	11.0
	28.2	33.5	43.9	48.0	42.4	50.3	66.0	72.2
Zod A	2.5	2.9	3.9	4.2	2.5	3.0	4.0	4.3
	11.4	13.6	17.8	19.5	18.1	21.5	28.2	30.8
Zod B	3.5	4.2	5.5	6.0	3.4	4.1	5.4	5.9
	14.3	17.0	22.3	24.4	19.8	23.5	30.8	33.7

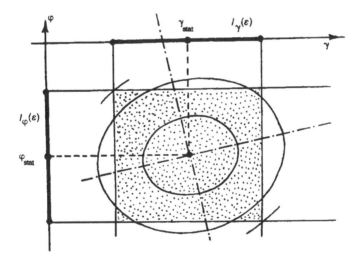

Figure 6.10. Determination of admissible variations of mean square latitudinal deviations. The line levels of $\sigma^2(\gamma, \varphi, t)$ and the confidence triangle for the estimates of (γ, φ) are shown.

tions of stars; the methods are in principle only applicable to large collections. As a result, we have found out that the groups A and $Zod\ A$ have the best accuracy and carry similar systematic errors. The groups C, D and M contain poorly measured stars and carry different systematic errors. Therefore, in the

Table 6.4.

		Zod A	A
a_{11}		1.11	0.82
a_{12}		0.042	−0.003
a_{22}		0.073	0.13
s_{min}		12″8	16″5
$\Delta\sigma$	$\varepsilon = 0.1$	1″3	1″2
	$\varepsilon = 0.05$	1″8	1″7
	$\varepsilon = 0.01$	3″0	1″8
	$\varepsilon = 0.005$	3″5	3″3
σ_{max}	$\varepsilon = 0.1$	14″1	17″7
	$\varepsilon = 0.05$	14″6	18″2
	$\varepsilon = 0.01$	15″8	19″3
	$\varepsilon = 0.005$	16″3	19″8

sequel we will not consider the stars of the last three groups. The stars in B and $Zod\ B$, though measured worse than the ones in A and $Zod\ A$, may carry the same systematic error; so far we cannot state this for sure.

The further analysis is due to the following problem. The parameters φ_{stat} and γ_{stat} that determine the systematic errors had been found from an analysis of a large group of stars, and have the sense of a turn of the ecliptic that minimizes the mean square latitudinal deviation for the stars in this collection. However, we cannot exclude a priori the possibility that *some minor subcollection of the large collection carries its own systematic error*, in which case the above parameters φ_{stat} and γ_{stat} are nothing more than the result of averaging true systematic errors, and are of little use for us.

Note that the lengths of confidence intervals for φ_{stat} found in the previous section are very large. This may be explained by insensitivity of latitudinal deviation to turns through the angle φ, as well as by the error in φ not being systematic. The different behavior of the two parameters may be easily explained if we consider, for example, the armillary sphere (see Chapter 1). In measurements with the help of this instrument, inclination of the ecliptic to the equator is fixed, so an error in this fixing affects coordinates of all stars measured with the help of this particular instrument. As for the error in φ, it is of a different nature: it arises in each individual measurement, and varies from star to star.

Therefore, it seems relevant to find the group errors for each constellation and compare them with the systematic error in the best measured collection $Zod\ A$.

2. Below we investigate twenty-one minor collections of stars, enlisted in Table 6.5. The structure of this table is similar to that of Table 6.1. We have selected for study the zodiacal constellations and vicinities of all named stars

Table 6.5.

Group of stars	Baily's numbers	Number of stars
Zodiacal constellations		
Aries	362–371, 373, 374	12
Taurus	380–388, 390, 391, 393–410	29
Gemini	424–440	17
Cancer	449–454	6
Leo	462–481, 483–488	26
Virgo	497–516, 518–520	23
Libra	529–534	6
Scorpius	546–565	20
Sagittarius	570–573, 575–583, 585, 586, 590, 591, 593, 594, 596–598	22
Capricornus	601–608, 610–627	26
Aquarius	629–650, 652–656, 658–660, 662–668	37
Pisces	674–695, 697, 699–701, 704–706	29
Vicinities of named stars		
Arcturus	88–96, 98, 100– 110	21
Antares	546–569	24
Aquila	286–300	15
Capella	220–233	14
Regulus	462–481, 483–488, 491–493	29
Sirius	812, 818–835, 837–846	29
Spica	497–503, 505–515, 518–526	27
Vega	149–158	10
Aselli	449–454, 456–461	12

except Canopus and Vindemiatrix (for the reasons we have explained above) and Procyon (whose neighborhood contains too few stars).

Finding group errors for individual constellations encounters some difficulties. Suppose G is a constellation, and φ_{stat}^{G} and γ_{stat}^{G} are the values found from the method of least squares, σ_{min} is the minimum possible residual mean square latitudinal deviation, and p_{stat}^{G} is the rate of stars whose residual latitudinal deviation is less than $10'$ (at $t = 18$). Because each individual constellation contains comparatively few stars, the statistical errors in γ_{stat}^{G} and φ_{stat}^{G} are too large for any reliable conclusions. Therefore, we also calculate σ_{1}^{G} and p_{1}^{G}, the mean square latitudinal deviation and the rate of stars whose latitudinal deviation is less than $10'$ (for $t = 18$) under the condition that the pole of the ecliptic is at the position appropriate to $Zod\ A$ (in different words, under the condition that the group errors are equal to $\gamma_{stat}^{Zod\ A}$ and $\varphi_{stat}^{Zod\ A}$).

If σ_{1}^{G} is but a little greater than σ_{min}^{G}, then we have the right to conclude that the group error in G is similar to that in $Zod\ A$. The difference between p_{1}^{G} and p_{min}^{G} is another criterion of similarity of the group error and the systematic error. Recall that σ_{min}^{G} and σ_{1}^{G} do not depend on t if we consider a collection

Table 6.6.

Group of stars		σ^G_{init}	σ^G_{min}	σ^G_1	p^G_{init}	p^G_{min}	p^G_1
Zodiacal constellations							
Aries	Z1	19.7	17.2	18.9	45.5%	45.5%	72.7%
Taurus	Z2	23.2	18.1	20.6	27.6%	41.4%	41.4%
Gemini	Z3	17.8	10.5	11.0	29.4%	82.4%	58.8%
Cancer	Z4	13.8	4.3	5.2	33.3%	100%	100%
Leo	Z5	20.2	11.1	11.2	19.2%	65.4%	65.4%
Virgo	Z6	18.4	13.6	14.4	39.1%	56.5%	47.8%
Libra	Z7	8.4	6.1	9.3	83.3%	83.3%	83.3%
Scorpius	Z8	18.8	13.7	15.1	30.0%	65.0%	55.0%
Sagittarius	Z9	16.4	14.3	15.8	30.4%	60.9%	60.9%
Capricornus	Z10	16.2	10.6	11.3	42.3%	65.4%	57.7%
Aquarius	Z11	28.6	17.3	19.2	18.4%	44.7%	44.7%
Pisces	Z12	22.5	21.5	21.7	51.7%	41.4%	34.5%
Vicinities of named stars							
Antares	S1	17.7	12.6	13.8	33.3%	70.8%	58.3%
Aselli	S2	15.7	11.0	12.1	33.3%	58.3%	66.7%
Capella	S3	34.6	30.3	34.0	35.7%	14.3%	64.3%
Aquila	S4	24.0	23.7	26.7	40.0%	33.3%	13.3%
Vega	S5	20.0	14.1	17.1	50.0%	60.0%	30.0%
Arcturus	S6	24.2	17.2	20.0	19.0%	38.1%	28.5%
Sirius	S7	15.2	11.9	25.9	47.4%	52.6%	15.8%
Spica	S8	17.9	14.1	14.5	44.4%	48.1%	48.1%
Regulus	S9	25.2	21.0	21.1	17.2%	58.6%	58.6%

of fixed stars, and vary but slightly with time otherwise. A similar statement is true for the rate of the stars whose latitudinal deviations do not exceed 10′.

Table 6.6 displays the relevant numeric data, also presented in visual form in Figures 6.11 and 6.12. In Figure 6.11 we present data for constellations (denoted by Z1–Z12), and in Figure 6.12, for vicinities of named stars (S1–S9). It should be noted that although some named stars are zodiacal, their vicinities do not coincide with the corresponding zodiacal constellations, but are groups of stars of the constellations that have acquired names in Bayer's notation. These stars are the brightest, and are usually most reliably identified, which enhances reliability of the conclusions.

3. It follows from the data in Table 6.6 and the graphs in Figures 6.11 and 6.12 that zodiacal constellations in the domain A (Gemini, Cancer, Leo, Virgo, Libra, Scorpio) have a remarkable property that the mean square error σ_1 and the rate of stars whose latitudinal deviations do not exceed 10′ obtained under the assumption that the group error is equal to $\gamma^{Zod\ A}_{stat}$ and $\varphi^{Zod\ A}_{stat}$ differ but slightly from σ_{min} and p_{min}, that ensue under the assumption that the position of the pole of the ecliptic is "optimal" for the particular constellation.

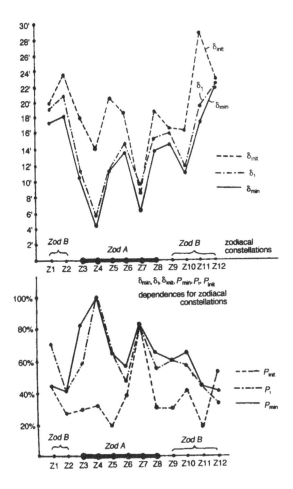

Figure 6.11. Analysis of group errors in zodiacal constellations. The graphs of the initial and residual (after compensation for the systematic errors) are shown (upper graphs). Here σ_{init} is the mean square error over the constellation before the compensation, σ_1 is the mean square error after compensation for the systematic error relevant for *Zod A*, and σ_{min} is the mean quadratic error after compensation for the systematic error found for this constellation (the minimum possible value). The rates of stars whose latitudinal deviations do not exceed $10'$, p_{init}, p_1, p_{min} (the lower graphs below).

The difference is greatest in the "best measured" constellation Libra, where all quantities σ_{init}, σ_{min} and σ_1 do not exceed $10'$, and $p_{init} = p_{min} = p_1 = 83.3\%$ (the rate of stars that have at most $10'$ latitudinal deviation). The equality $p_{init} = p_{min} = p_1 = 83.3\%$ may be explained by the fact that the constellation almost lies in the equinoctial axis, so the turn γ practically does not affect it.

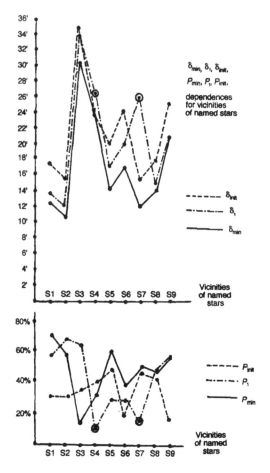

Figure 6.12. Analysis of systematic errors in vicinities of named stars. Notation is similar to that in Figure 6.11.

A similar remark applies to constellations in *Zod B*, although with some reservations; in fact, it is not very important for us, because *Zod B* contains no named stars. Nonetheless, we should note an interesting fact about Aries: although σ_1 is but a little less than σ_{init} (note that σ_{min} also differs but slightly from σ_{init}), we have $p_1 \gg p_{\text{init}} = p_{\text{min}}$; hence the displacement of the pole of the ecliptic to the position appropriate to *Zod A* sharply increases the rate of well-measured stars in Aries (up to 72.7%).

The following general conclusion may be drawn from the analysis of zodiacal constellations. If the "optimal" deviation σ_{min} is much less than σ_{init}, then the assumption that the group error is equal to the systematic error (in $Zod\ A$) and the corresponding compensation for this error leads to $\sigma_1 \ll \sigma_{init}$. This is true for Gemini, Cancer, Leo, Virgo, Scorpio, Capricornus and Aquarius.

If σ_{min} is close to σ_{init}, then as a rule, $\sigma_{min} \leq \sigma_1 \leq \sigma_{init}$, and the effect of displacement of the pole of the ecliptic to the position appropriate to $Zod\ A$ is barely noticeable. This property is characteristic for Aries (as we noted, the displacement sharply raises the rate of well-measured stars in Aries), Taurus, Libra, Sagittarius, and Pisces. Among these, the good accuracy properties of Libra are practically unaffected by the displacement of the pole of the ecliptic from the position "optimal" for the constellation to the position appropriate to $Zod\ A$, the accuracy of Aries even improves, and the properties of the rest of the constellation do not alter, remaining "average". A typical example is Taurus, for which $\sigma_{init} = 23\rlap.{''}2$, $\sigma_{min} = 18\rlap.{''}1$, $\sigma_1 = 20\rlap.{''}6$, $p_{init} = 27.6\%$, $p_{min} = p_1 = 41.4\%$. Of all constellations, Pisces stands out: here we have both $p_{min} < p_{init}$ and $p_1 < p_{init}$, and $\sigma_{init} \approx \sigma_{min} \approx \sigma_1$.

4.   The properties of vicinities of named stars vary still more. First of all, let us note the vicinities of Aquila and Sirius. In both cases, compensation for the systematic error appropriate to $Zod\ A$ brings about both an increase of the mean square latitudinal deviation (in the case of Sirius, an essential increase from $15\rlap.{''}2$ to $25\rlap.{''}9$) and a decrease of the rate of well-measured stars (for Aquila, from 40% to 13.3%, and for Sirius, from 47.4% to 15.8%). This means that the group errors appropriate to these vicinities differ from the one appropriate to $Zod\ A$; unfortunately, a reliable determination of these group errors seems impossible. *Therefore we exclude neighborhoods of Sirius and Aquila from our further treatment.*

The properties of the rest of the vicinities of named stars are quite similar to that of zodiacal constellations. Namely, compensation for the group error appropriate to $Zod\ A$ essentially decreases (down to the values quite close to the minimum possible) the mean square error for the vicinities of Antares, Aselli, Arcturus, Spica and Regulus; the rates of stars whose latitudinal deviations do not exceed $10'$ also increases. The vicinity of Capella has properties similar to that of Aries: the mean square latitudinal deviation is practically unaffected by the displacement of the observer's pole of the ecliptic to the position computed for $Zod\ A$, but the displacement raises to 64.3% the rate of stars with individual latitudinal deviations below $10'$ (the rate being equal to 36.7% initially, and 14.3% in the "optimal" case!). The picture is quite the opposite in the vicinity of Vega: the displacement improves essentially the mean square latitudinal deviation, but decreases the rate of stars with small individual latitudinal deviations. Thus, the character of group errors in the vicinities of Capella and Vega remains unclear.

5.   Although we have shown the similarity of the characteristics σ_1 and p_1 to σ_{min} and p_{min} for most constellations, the question whether the error

Table 6.7.

Group of stars		σ_{init}^G	σ_{min}^G	σ_2^G	p_{init}^G	p_{min}^G	p_2^G
Aries	Z1	19.7	17.2	17.2	45.5%	45.5%	45.5%
Taurus	Z2	23.2	18.1	20.2	27.6%	41.4%	41.4%
Gemini	Z3	17.8	10.5	10.6	29.4%	82.4%	82.4%
Cancer	Z4	13.8	4.3	4.5	33.3%	100%	100%
Leo	Z5	20.2	11.1	11.1	19.2%	65.4%	65.4%
Virgo	Z6	18.4	13.6	14.4	39.1%	56.5%	52.2%
Libra	Z7	8.4	6.1	6.1	83.3%	83.3%	83.3%
Scorpius	Z8	18.8	13.7	13.7	30.0%	65.0%	70.0%
Sagittarius	Z9	16.4	14.3	14.4	30.4%	60.9%	56.5%
Capricornus	Z10	16.2	10.6	10.6	42.3%	65.4%	65.4%
Aquarius	Z11	28.6	17.3	18.7	18.4%	44.7%	47.4%
Pisces	Z12	22.5	21.5	21.7	51.7%	41.4%	37.9%

φ_{stat} is systematic remains open. In order to answer it, let us find for zodiacal constellations (that contain also the six named stars) two more characteristics. Consider all possible errors determined by $\gamma = \gamma_{stat}^{Zod\ A}$ and various φ and find the value $\varphi^{(2)}$ of φ that minimizes the residual mean square deviation over the constellation after compensation for the error; we denote the minimum value of the mean square deviation by σ_2, and the corresponding rate of stars with individual deviations below 10′ by p_2. Thus, for a constellation G,

(1)
$$\sigma_2^G = \sigma_2^G(t) = \min_{\varphi} \sigma^G(\gamma_{stat}^{Zod\ A}, \varphi, t)$$

(2)
$$\varphi^{(2)} = \arg\min_{\varphi} \sigma^G(\gamma_{stat}^{Zod\ A}, \varphi, t)$$

Table 6.7 gives the values of σ_2 and p_2; it is similar to Table 6.6, moreover, for reader's convenience we give here some data once more. The data are also represented graphically in Figure 6.13 in the fashion similar to that of Figure 6.11. It is obvious from the table and the figure that fixing $\gamma = \gamma_{stat}^{Zod\ A}$ and varying φ allows reaching the values of σ_2 equal to, or close to σ_{min}. The corresponding values of p_2 are also close to p_{min}. It is interesting that the picture for constellations in *Zod B* is similar.

This shows that $\gamma_{stat}^{Zod\ A}$ is really the systematic error made by the compiler of the catalog as he measured the stars in *Zod A* and the named stars (except Sirius, Aquila and Canopus). As for $\varphi_{stat}^{Zod\ A}$, it is apparently the result of averaging individual errors of measurement, and there are no grounds for treating it as a systematic error. Furthermore, the value of φ_{stat} is determined but roughly, so it provides little information.

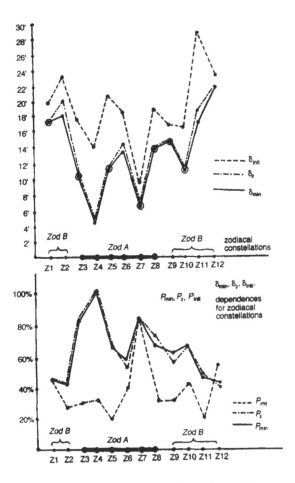

Figure 6.13. Analysis of stability of φ_{stat}. Notation is as in Figures 6.11 and 6.12, but in all cases the optimal value of γ for the constellation is chosen.

4. Brief conclusions

1. The above analysis confirms that the "observer's poles of the ecliptic" of the stars in A and Zod A are near, that is, the two domains are measured with the same systematic error.

2. The analysis provides no grounds to assume that systematic errors in the domains C, D and M coincide with the one for A and Zod A.

3. The accuracy of measurement of stars in A and Zod A is much higher than that in the rest of the domains.

4.   The residual mean square deviation in *Zod A* is 12′.8; about two-thirds of all stars in the domain have individual latitudinal deviations below 10′ (the claimed accuracy of the catalog). The similar characteristics of the domain *A* are 16′.5 and 1/2.

5.   An analysis of zodiacal constellations and vicinities of individual stars leads to the conclusion that the parameter γ (the error in the inclination of the ecliptic) is a systematic error. As for the parameter φ, it may be the result of averaging various groups or systematic errors.

6.   For Gemini, Cancer, Leo, Virgo, Libra, Scorpio, Sagittarius, Capricornus, and for vicinities of Antares, Aselli, Arcturus, Spica and Regulus, the group errors γ are close to (or coincide with) the systematic error γ_{stat} computed for the best measured domain *Zod A*.

7.   Nothing definite can be said about the group errors for Aries and Taurus, as well as for vicinities of Capella and Vega; the group errors for these groups of stars may coincide with or differ from the ones for *Zod A*.

8.   The group errors for vicinities of Sirius and Aquila differ from the one appropriate to *Zod A*, but no particular values for these errors can be found. The group error for Pisces is also likely to be different from $\gamma_{stat}^{Zod\ A}$.

Chapter 7

Dating the Star Catalog
of the *Almagest*

1. Informative kernel of the catalog

The aim of the analysis carried out in Chapters 1–6 was to diminish latitudinal errors in the coordinates of stars given in the catalog of the *Almagest*. Several complementary media was used to that end: exclusion of doubtfully identified stars, exclusion of outlies, finding group errors for separate parts of the catalog and compensation for the group errors, and confinement to the best measured parts of the catalog (after compensation for the group errors); as we have found out, the best measured is the domain A, and especially its part *Zod A*. After compensation for the group error, the rate of stars with latitudinal deviation below $10'$ is 50% in A and 64% in *Zod A*. We also analyzed individual constellations and have distinguished the ones in which the group error coincides with the systematic error characteristic for *Zod A*.

Thus, *we have confirmed that the accuracy* $10'$, *claimed by the compiler of the catalog, is really reached at least for most stars in the domain A.*

However, dating the catalog requires considering sufficiently fast and sufficiently well measured stars, so we need estimates for individual errors. The above statistical characteristics provide no information about which particular stars are measured well and which are not.

A choice of such stars can only be based on some plausible assumptions founded on our knowledge of practical rules of measuring stellar positions (see Chapter 1).

In antiquity and the Middle Ages, and in fact nowadays the so-called *reference stars* were used, which are few in comparison with the total number of stars in the catalog. It is known that Tycho Brahe used 21 reference stars, disposed in the zodiac[34]. The modern system of reference stars comprises several thousands stars[29]. Unfortunately, we do not know what stars were used in the *Almagest* for reference; it is only clear that Regulus and Spica had to be among them, because particular sections of the *Almagest* are devoted to their measurement. It is natural to suppose that the compiler of the catalog measured especially thoroughly the named stars; there are twelve named stars in the catalog: Arcturus (α Boo, 110), Regulus (α Leo, 469), Spica (α Vir, 509), Vindemiatrix (ε Vir, 509), Capella (α Aur, 222), Lyra (Vega; α Lyr, 149), Procyon (α CMi, 848), Sirius (α CMa, 818), Antares (α Sco, 553), Aquila (Altair; α Aq, 288), Aselli (γ Can, 452) and Canopus (α Argo Navis, 892). All these stars are bright and stand out sharply against their vicinities, and they form a convenient base in the celestial sphere. Many of these stars have a notable proper motion, and some (Arcturus, Procyon and Sirius) have high velocities of proper motion. Seven of the stars lie either in *Zod A* or immediately near it (Arcturus, Spica, Procyon, Aselli, Vindemiatrix, Regulus and Antares), and nine border the domain *A* (add Lyra and Vega to the seven stars). Thus, even if the twelve stars did not serve as references, they were to be measured especially thoroughly.

However, the twelve stars do not all have equal status.

First, Canopus is too much to the south, so refraction greatly affects its apparent position; hence its coordinates in the catalog are given with too large an error (greater than 1°).

Second, the coordinates of Vindemiatrix originally measured by the compiler of the catalog are not known; only the results of later measurements are available[22].

Furthermore, the group errors in measurements of vicinities of Sirius and Aquila differ from the group errors for the rest of the stars found in Chapter 6. Since we cannot determine the group errors for these two stars, we cannot compensate for them.

Thus, eight named stars are left for use in dating, whose vicinities have the same group error (at least, the same component γ of the group error). We will call the collection of the eight stars the *informative kernel*.

It is natural to assume that *since the claimed accuracy of the catalog is confirmed, it must be attained at the stars in the informative kernel (after compensation for the group error)*. Our further considerations will be based on this assumption.

Meanwhile, it is far from obvious that any date for the catalog may be obtained from an analysis of the informative kernel. Although we have restored the true accuracy of the catalog by compensating for the group errors, it is not at once clear that the residual error is sufficiently small. Furthermore, although we have proved similarity of the group errors for the vicinities of stars in the informative kernel, this does not necessarily mean that the individual errors for the stars are the same. Although the situation that the central star is measured with an essentially different error in comparison with its vicinity seems unnatural, strictly speaking we cannot a priori reject this possibility. We also should not exclude the possibility that a star in the informative kernel is measured to a worse accuracy than 10'.

Thus, the very existence of an interval of time for which the above assumption is true is an additional confirmation for the statistically derived conclusions.

2. Preliminary considerations

1. In the previous section we have distinguished a group of stars, which we called the *informative kernel* of the *Almagest*; we will analyze in details its behavior in the following sections. In this section we will look at the behavior of the group consisting of all the twelve stars that are named in the *Almagest*. This analysis demonstrates the improvement of accuracy of the catalog due to compensation for the systematic error, and provides additional grounds for the statement that the three stars of twelve (Canopus, Sirius and Aquila) break the homogeneity of the sampling, being outlies in relation to the rest of the named stars (a reason for this was given in Chapter 6, on the basis of an analysis of group errors in the latitudes of stars in various domains of the sky).

Denote by $\Delta B_i(t, \gamma, \varphi)$ the difference between the latitude of the ith named star of the *Almagest* ($1 \leq i \leq 12$) obtained after compensation for the group error (γ, φ) and its true latitude computed for the epoch t.

2. Let us look at how the true accuracy of latitudes relates to the value of scale division (10') *under the assumption that the catalog contains no global* systematic errors. Table 7.1 exposes absolute latitudinal deviations of the twelve named stars for various a priori dates t. The first column of the table contains Baily numbers of the stars; the values of latitudinal deviations are given in minutes of arc.

It is obvious from the table that for seven of the twelve stars, the latitudinal deviations exceed 10' for all possible values of t. In the columns corresponding to traditional dates 100 AD (Ptolemy's time) and 200 BC (Hipparchus' time) the large error in the coordinate of Arcturus (30'–40') stands out. It seems strange that the brightest star in the Northern hemisphere was observed by Ptolemy (or Hipparchus) worst. Further, it follows from the text of the *Almagest* that the coordinates of Regulus had been measured several times and that this star served as one of the reference points for measurement of coordinates of other stars, so it is natural to expect that Regulus had been measured with highest accuracy, so its latitudinal deviation should not

Table 7.1.

	1800 AD	1400 AD	900 AD	400 AD	100 AD	200 BC
Arcturus (110)	37.8	21.2	0.9	19.3	31.4	43.3
Sirius (818)	23.6	18.3	11.7	5.1	1.2	2.6
Aquila (288)	8.6	9.4	10.5	11.8	12.6	13.4
Vindemiatrix (509)	13.0	14.3	15.8	17.1	17.8	18.4
Antares (553)	32.6	29.5	25.5	21.6	19.3	17.0
Aselli (452)	30.5	28.5	25.9	23.2	21.5	19.8
Procyon (848)	11.2	16.0	21.9	27.6	31.1	34.4
Regulus (469)	17.5	16.6	15.4	14.0	13.0	12.1
Spica (510)	2.4	0.7	1.3	3.1	4.2	5.2
Lyra (149)	15.4	14.2	12.5	10.8	9.8	8.7
Capella (222)	21.9	21.7	21.3	21.0	20.8	20.6
Canopus (892)	51.0	54.2	58.2	62.3	64.8	67.3

Table 7.2.

	1800 AD	1400 AD	900 AD	400 AD	100 AD	200 BC
Arcturus (110)	29.9	15.5	2.3	20.0	30.5	41.0
Sirius (818)	44.2	39.2	32.7	25.9	21.8	17.5
Aquila (288)	27.0	28.7	30.7	32.5	33.5	34.4
Vindemiatrix (509)	15.6	14.9	13.8	12.6	11.8	11.0
Antares (553)	13.3	11.0	8.5	6.2	4.9	3.7
Aselli (452)	13.2	10.2	6.5	2.9	0.9	1.1
Procyon (848)	8.1	4.0	1.2	6.7	10.1	13.5
Regulus (469)	6.1	3.5	0.4	2.7	5.1	6.2
Spica (510)	5.1	4.9	4.4	3.7	3.3	2.7
Lyra (149)	5.1	6.7	8.5	10.0	10.8	11.5
Capella (222)	1.3	1.5	2.1	2.9	3.5	4.2
Canopus (892)	71.5	75.0	79.2	83.1	85.4	87.6

exceed 10'. Note that the latitudinal deviation of another bright star in the ecliptic, Spica, whose coordinates had also been measured by Ptolemy separately and which also was then used as a reference star (see Chapter VII.2 of the *Almagest* [17]), does not exceed 5', half the value of the least division of the catalog.

3. Let us now take into account the systematic error found in Chapter 6. Since the component γ varies slowly with time from 1 AD to the Middle Ages, and φ affects the picture but little, we will set $\gamma = 21'$ and $\varphi = 0$.

Table 7.2 exposes latitudinal deviations of stars computed after compensation for the systematic error determined by $\gamma = 21'$ and $\varphi = 0$.

A comparison of Tables 7.1 and 7.2 shows that compensation for the systematic error sharply improves accuracy of coordinates of the named stars for all a priori dates. The latitudes of Regulus and Spica here turn out to be measured to an accuracy within $5'$ whatever be the date of compilation of the catalog from late antiquity to late Middle Ages. Furthermore, for the dates in the interval $6 \le t \le 10$ (900 AD–1400 AD), the latitudinal deviations do not exceed $10'$ for eight of the twelve named stars. The eight stars are exactly the ones contained in the domain A we have distinguished in Chapter 6 as we carried out a statistical analysis of the catalog of the *Almagest*.

Naturally, the above argument requires a further justification; in particular, other values of γ and φ should be tried. Detailed calculations and more exact statements are given in subsequent sections of this chapter.

4. An additional information about the date of compilation of the catalog may be obtained from the following. Let E be a collection of stars in the *Almagest* and $\Delta B_i(t, \gamma, \varphi)$ be the latitudinal deviation of the ith star in the collection ($1 \le i \le n$). Let us construct empirical distribution functions for latitudinal errors for the stars in E:

$$(1) \qquad F_{t,\gamma,\varphi}(x) = \frac{1}{n}\#\{i : |\Delta B_i(t, \gamma, \varphi)| \le x\}$$

where n is the number of stars in E, and $\#\{\cdot\}$ denotes the number of stars in the set $\{\cdot\}$. Comparing these functions constructed for various values of t, γ and φ, we may try to select the values so that the latitudinal errors for the stars in E are minimal (in the stochastic sense). As a measure of distinction between two sets of errors, we will use the mean difference. This may be represented visually as the area of the domain between the graphs of the two distribution functions; the area of each of the pieces between the graphs is taken with the sign depending on which graph bounds the domain from the right and which from the left (see Figure 7.1). The distribution function $F_{t_0,\gamma_0,\varphi_0}$ that is more to the left than the rest of the functions $F_{t,\gamma,\varphi}$ corresponds to the least errors in latitudes of stars in E. In this case, it is natural to treat the date t_0 and the value of the systematic error (γ_0, φ_0) as approximations of the true date and the true systematic error.

Let us use the well-known catalog of Tycho Brahe, compiled in the second half of the 16th century to illustrate this. We considered the set E of thirteen named stars in the catalog, and have constructed the empiric distribution functions $F_{t,\gamma,\varphi}$ for $\gamma = \varphi = 0$ and three values of t: $t = 3$ (1600 AD), $t = 3.5$ (1550 AD) and $t = 4$ (1500 AD). The result is displayed in Figure 7.2. It is obvious from the figure that if we do not take into account a possible systematic error (we have put $\gamma = \varphi = 0$), the optimal date for the catalog is $t = 3.5$ (1550 AD); this date minimizes in the above sense the errors in latitudes of the thirteen named stars. The date 1550 AD is really near the true date of compilation of the catalog. The thirteen named stars are Regulus, Spica, Arcturus, Procyon, Lyra (Vega), Sirius, Capella, Aquila and Antares,

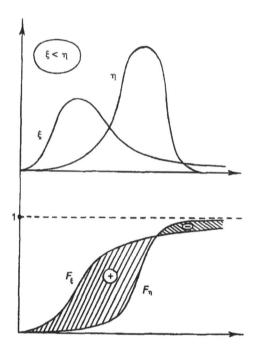

Figure 7.1. Comparison of random variables. The area marked by plus can be viewed as a measure that ξ is greater than η.

named also in the *Almagest*, and also Caph (β Cas), Denebola (β Leo), Pollux (β Gem) and Shiat (β Peg).

Let us now consider the empiric functions of distribution $F_{t,\gamma,\varphi}$ for the set E of the twelve named stars of the *Almagest* (see Section 1). In Figure 7.3 we give the graphs of these functions for various γ and $t = 5$ (1400 AD), $t = 10$ (900 AD), $t = 18$ (100 AD), and $t = 20$ (100 BC); we have set $\varphi = 0$, because the picture is but little sensitive to the value of φ. The values $t = 10$ and $\gamma = 21'$ turn out to be optimal.

The behavior of the graphs of $F_{t,\gamma,\varphi}$ for the *Almagest* is a little sensitive to variations of the choice of the named stars. For comparison, we give in Figure 7.4 the graphs for the thirteen stars that are named in the catalog of Tycho Brahe (with the coordinates as given in the *Almagest*); the values $\gamma = 21'$ and $t = 10$ are optimal for this set of stars too. In Figure 7.4, the difference between the values $\gamma = 0$ and $\gamma = 21'$ is quite clear: all the graphs corresponding to $\gamma = 21'$ are more to the left (and so the errors are smaller) than the ones corresponding to $\gamma = 0$. In other words, the estimate $\gamma = 21'$ is "better" than $\gamma = 0$ for all t in the a priori dating interval.

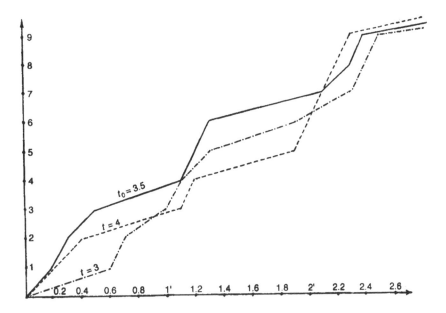

Figure 7.2. Comparison of distribution functions of latitudinal deviations in Tycho Brahe's catalog at various a priori dates. Obviously, the date $t_0 = 3.5$ (1550 AD) is optimal.

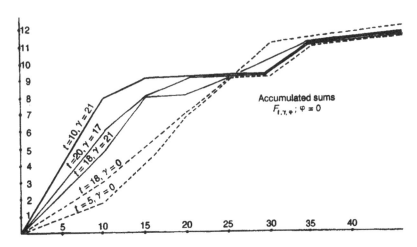

Figure 7.3. Comparison of distribution functions of latitudinal deviations in the collection of the twelve named stars of the *Almagest* at various a priori dates and various compensations for the systematic error. Obviously, the optimal date is $t = 10$ (900 AD).

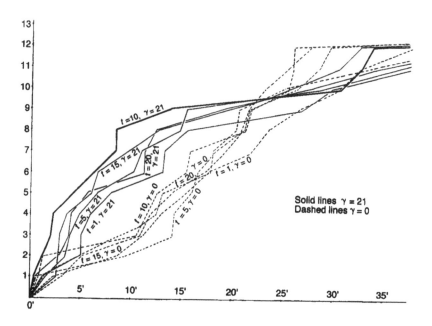

Figure 7.4. Comparison of distribution functions of latitudinal deviations in the collection of the thirteen named stars of Tycho Brahe's catalog mentioned in the *Almagest* at various a priori dates and various compensations for the systematic error. Obviously, the optimal date is $t = 10$ (900 AD).

5. To conclude this section, let us discuss whether it is possible to expand the informative kernel of the *Almagest* and not worsen the latitudinal accuracy of the collection. It appears natural to start from the collection of all stars that have by now acquired proper names. The names were given to stars in antiquity, and possibly the author of the *Almagest* attributed special significance to many of them. The list of all such stars is given in Table Ap. 2 in the Appendix.

It turns out, however, that this expansion of the collection of named stars of the *Almagest* makes the accuracy of latitudes much worse. Figure 7.5 shows the dependence of the rates of stars with latitudinal errors below 10′ and 20′ (after compensation for the error with $\gamma = 20'$) on time. The graphs demonstrate that only 32 stars of 52 are ever within 20′ (in latitude), and only about one-third of all stars (19 of 52) are ever within 10′ from their positions given in the *Almagest*. The dashed line in Figure 7.5 is the graph of mean latitudinal deviation (after compensation for the systematic error with $\gamma = 20'$) of the collection of 38 stars whose latitudinal deviations are at some t below 10′. The graphs demonstrate that the mean deviation varies but slightly with t.

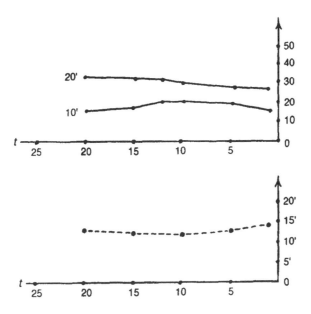

Figure 7.5. The collection of 52 stars that have medieval names and are mentioned in the *Almagest* (as a rule, they have no names in the *Almagest*). Upper graphs: The number of stars whose latitudinal deviations after compensation for the systematic error do not exceed 10′ (20′) dependence on the a priori date. Lower graph: The mean latitudinal deviation over the collection dependence of the a priori date. It is obvious that both characteristics depend but slightly on the a priori date, so the collection is of little use for dating purposes.

We have also considered different possibilities of expanding the informative kernel (for example, taking into account brightness of the stars); all attempts lead to a sharp worsening of accuracy and to the loss of sensitivity of characteristics of the expanded collection to the date *t*. Apparently, this worsening is due to the fact that the expansions of the informative kernel lead to the inclusion of stars that lie in the domains with different group errors.

3. Statistical dating procedure

1. The hypothesis that the real accuracy of measurement of named stars in the catalog corresponds to the claimed accuracy gave us the possibility to give in Section 2 a *qualitative* answer about the date of compilation. Namely, we have shown that the configuration of the informative kernel varies sufficiently fast to provide a possibility to determine the date. So now we may consider the question about the *quantitative* determination of a dating interval.

The following procedure, which we call *statistical*, looks natural within the framework of our approach. The procedure uses, on the one hand, the hypothesis about the accuracy of measurement of named stars, and on the other hand, the statistical characteristics of group errors found in Chapter 6.

Statistical Dating Procedure

a) Fix the confidence level $1 - \varepsilon$.

b) Fix a moment of time t and find the confidence interval $I_y(\varepsilon)$ for the corresponding value of the component $\gamma_{\text{stat}}^{\text{Zod } A}$ of the group error for *Zod A*. Put

$$(1) \qquad\qquad \Delta(t) = \min \Delta(t, \gamma, \varphi)$$

where the minimum is taken over all $\gamma \in I_y(\varepsilon)$ and all possible values of φ, and $\Delta(t, \gamma, \varphi) = \max_{1 \le i \le 8} |\Delta B(t, \gamma, \varphi)|$ is the maximum latitudinal deviation over the stars in the informative kernel computed for the moment t after compensation for the systematic error (γ, φ).

c) If $\Delta(t)$ does not exceed $10'$, the claimed accuracy of the catalog, then the year t should be treated as a possible date of compilation of the catalog; otherwise the catalog cannot be attributed to the year t.

Of course, the result of application of the statistical procedure depends on the choice of the confidence level $1 - \varepsilon$. Therefore the ensuing results are to be tested for stability with respect to variations of ε. This, as well as some other tests will be carried out below.

2. Before we start to present our final results, we first give some information on how $\Delta(t, \gamma, \varphi)$ depends on t, γ and φ. In order not to overload our exposition, we represent the dependence in graphical form; see Figures 7.6–7.23. Each picture corresponds to a moment of time $t = 1, \ldots, 18$ (recall that $t = 1$ is 1800 AD, and $t = 18$ is 1 AD). The values of γ are plotted as abscissas and the values of φ as ordinates; the domains where $\Delta(t, \gamma, \varphi) \le 10'$ are crosshatched, the domains where $10' < D(t, \gamma, \varphi) \le 15'$ are hatched, and the domains where $15' < D(t, \gamma, \varphi) \le 20'$ are dotted; in blank areas, $\Delta(t, \gamma, \varphi) > 20'$. The bold dot in each figure marks the point $(\gamma_{\text{stat}}^{\text{Zod } A}(t), \varphi_{\text{stat}}^{\text{Zod } A}(t))$. It is obvious from the figures that the crosshatched domains only appear at $6 \le t \le 13$, and the hatched domains at $4 \le t \le 16$; the spots attain maximum size at $7 \le t \le 12$. At $t \ge 18$ (note that these values of t cover both the epochs of Ptolemy and Hipparchus), no points with $\Delta(t, \gamma, \varphi) \le 20'$ exist in the area displayed. In other words, the maximum latitudinal deviation of the named stars that corresponds to attributing the catalog to 100 AD or earlier is at least two times as large as the claimed accuracy of the catalog. In fact, this error even exceeds the residual mean square

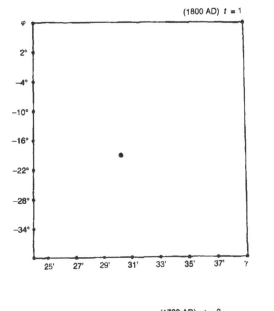

Figure 7.6.

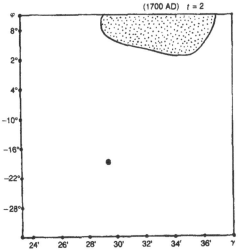

Figure 7.7.

Figures 7.6 through 7.23. The minimax deviation $\Delta(t, \gamma, \varphi)$ dependence on the a priori date t and the parameters of the systematic error γ and φ. At fixed t the rectangular confidence neighborhood of the point $(\gamma_{stat}^{Zod}\ ^A(t), \varphi_{stat}^{Zod}\ ^A(t))$ (marked bold) in the plane with the coordinates (γ, φ) is shown. The crosshatched domains are the sets of the parameters that lead to the minimax latitudinal deviation below 10'. The domains where the deviation is below 15' are hatched, and those with the deviation below 20' are dotted. The values of t for which the crosshatched domains are not void constitute the interval of admissible dates for the catalog. The figures cover the intervals 100–1800 AD.

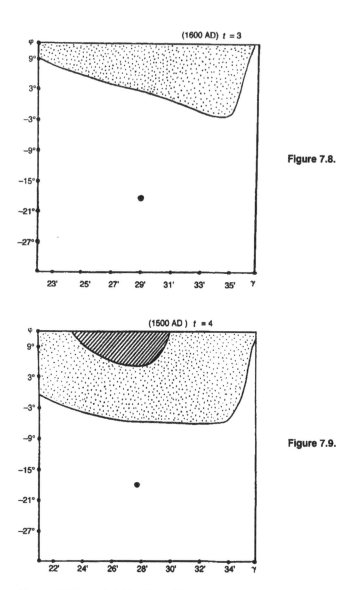

Figure 7.8.

Figure 7.9.

Figures 7.6 through 7.23. The minimax deviation $\Delta(t, \gamma, \varphi)$ dependence on the a priori date t and the parameters of the systematic error γ and φ. At fixed t the rectangular confidence neighborhood of the point $(\gamma_{\text{stat}}^{Zod}{}^{A}(t), \varphi_{\text{stat}}^{Zod}{}^{A}(t))$ (marked bold) in the plane with the coordinates (γ, φ) is shown. The crosshatched domains are the sets of the parameters that lead to the minimax latitudinal deviation below 10′. The domains where the deviation is below 15′ are hatched, and those with the deviation below 20′ are dotted. The values of t for which the crosshatched domains are not void constitute the interval of admissible dates for the catalog. The figures cover the intervals 100–1800 AD.

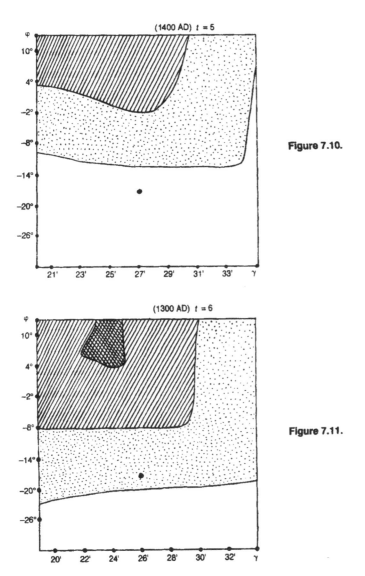

Figure 7.10.

Figure 7.11.

Figures 7.6 through 7.23. The minimax deviation $\Delta(t, \gamma, \varphi)$ dependence on the a priori date t and the parameters of the systematic error γ and φ. At fixed t the rectangular confidence neighborhood of the point $(\gamma_{stat}^{Zod\ A}(t), \varphi_{stat}^{Zod\ A}(t))$ (marked bold) in the plane with the coordinates (γ, φ) is shown. The crosshatched domains are the sets of the parameters that lead to the minimax latitudinal deviation below 10'. The domains where the deviation is below 15' are hatched, and those with the deviation below 20' are dotted. The values of t for which the crosshatched domains are not void constitute the interval of admissible dates for the catalog. The figures cover the intervals 100–1800 AD.

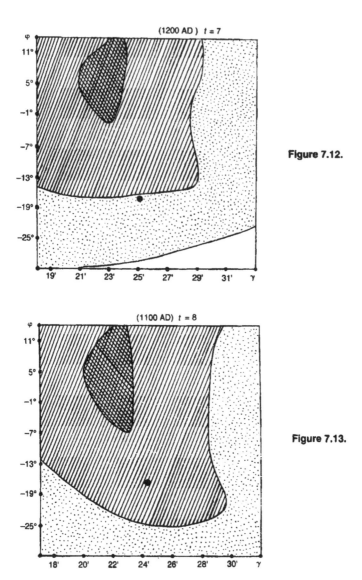

Figure 7.12.

Figure 7.13.

Figures 7.6 through 7.23. The minimax deviation $\Delta(t, \gamma, \varphi)$ dependence on the a priori date t and the parameters of the systematic error γ and φ. At fixed t the rectangular confidence neighborhood of the point $(\gamma_{\text{stat}}^{\text{Zod } A}(t), \varphi_{\text{stat}}^{\text{Zod } A}(t))$ (marked bold) in the plane with the coordinates (γ, φ) is shown. The crosshatched domains are the sets of the parameters that lead to the minimax latitudinal deviation below 10'. The domains where the deviation is below 15' are hatched, and those with the deviation below 20' are dotted. The values of t for which the crosshatched domains are not void constitute the interval of admissible dates for the catalog. The figures cover the intervals 100–1800 AD.

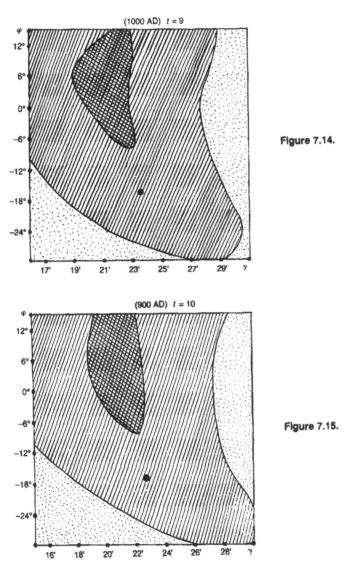

Figure 7.14.

Figure 7.15.

Figures 7.6 through 7.23. The minimax deviation $\Delta(t, \gamma, \varphi)$ dependence on the a priori date t and the parameters of the systematic error γ and φ. At fixed t the rectangular confidence neighborhood of the point $(\gamma_{stat}^{Zod\ A}(t), \varphi_{stat}^{Zod\ A}(t))$ (marked bold) in the plane with the coordinates (γ, φ) is shown. The crosshatched domains are the sets of the parameters that lead to the minimax latitudinal deviation below 10′. The domains where the deviation is below 15′ are hatched, and those with the deviation below 20′ are dotted. The values of t for which the crosshatched domains are not void constitute the interval of admissible dates for the catalog. The figures cover the intervals 100–1800 AD.

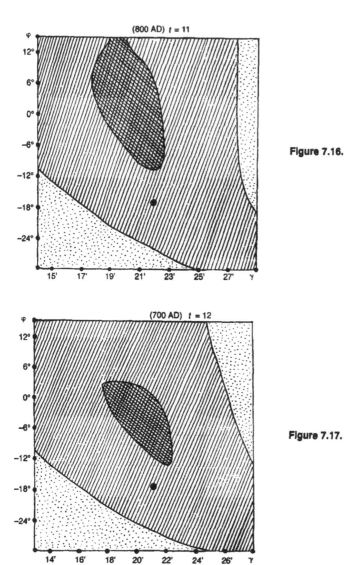

Figure 7.16.

Figure 7.17.

Figures 7.6 through 7.23. The minimax deviation $\Delta(t, \gamma, \varphi)$ dependence on the a priori date t and the parameters of the systematic error γ and φ. At fixed t the rectangular confidence neighborhood of the point $(\gamma_{\text{stat}}^{Zod\ A}(t), \varphi_{\text{stat}}^{Zod\ A}(t))$ (marked bold) in the plane with the coordinates (γ, φ) is shown. The crosshatched domains are the sets of the parameters that lead to the minimax latitudinal deviation below 10'. The domains where the deviation is below 15' are hatched, and those with the deviation below 20' are dotted. The values of t for which the crosshatched domains are not void constitute the interval of admissible dates for the catalog. The figures cover the intervals 100–1800 AD.

errors for the domains A, Zod A, B and Zod B, and is close to the residual mean square error for the domain M. Because of their emptiness, we do not adduce the pictures for $t \geq 18$.

Thus, within the ranges of γ and φ used in the figures (in fact, rather large), it is impossible to make the latitudinal deviations of all the eight named stars less than $10'$ if $t < 6$ (after 1300 AD) and if $t > 13$ (before 600 AD). If we raise the admissible error up to $15'$, then this level cannot be reached after 1500 AD and before 300 AD. We will discuss these facts in more details below, and now we will obtain a dating interval for the *Almagest* from the statistical procedure described as above.

3. Let us fix t and the level $\alpha > 0$, and put

(2) $$S_t(\alpha) = \{\, \gamma : \min_{\varphi} \Delta(t, \gamma, \varphi) \leq \alpha \,\}$$

Generally speaking, the set $S_t(\alpha)$ can be empty. Let us choose $\alpha = 10'$ and consider the intersection of $S_t(\alpha)$ with the confidence interval $I_y(\varepsilon)$ about $\gamma_{\text{stat}}^{Zod\ A}(t)$. If the intersection is nonempty, then in correspondence with the statistical dating procedure, we declare the moment t to be a possible date of compilation of the catalog. The set of all such t is the *interval of admissible dates*.

Figure 7.24 visualizes this construction. The domain $\{\, (t, \gamma) : \gamma \in S_t(\alpha) \,\}$ with $\alpha = 10'$ is dotted, and the similar domain with $\alpha = 15'$ is blank (we will use it below). The graph of $\gamma_{\text{stat}}^{Zod\ A}(t)$ we have constructed in Chapter 6 (see Figure 6.8), and the lengths of confidence intervals $I_y(\varepsilon)$ may be found in Table 6.3.

It follows from Figure 7.24 that *for all $\varepsilon \leq 0.1$, the interval of admissible dates is the same, $6 \leq t \leq 13$ (600 AD–1300 AD)*.

4. The comparatively large length of the ensuing interval may be explained in several ways. First of all, the accuracy of the catalog is comparatively low: the latitude of the fastest among the eight stars of the informative kernel (Arcturus) varies by the $10'$ in approximately 260 years.

Another reason is that in our calculations we only used confidence intervals for the component γ of the group error, minimizing $\Delta(t, \gamma, \varphi)$ over *all possible* values of φ (see (1) and (2)). Clearly, this approach leads to an expansion of the interval of admissible dates. Indeed, if we could think of φ as a group error, then we would have to choose the value for φ from its confidence interval, which would raise the value of $\min_{\varphi} D(t, \gamma, \varphi)$, hence would narrow the interval of admissible dates. However, we have no grounds to treat φ as a group error in the collections of stars in question.

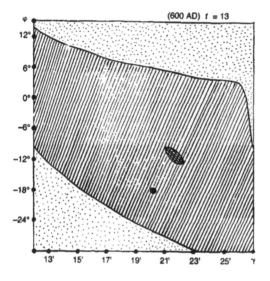

Figure 7.18.

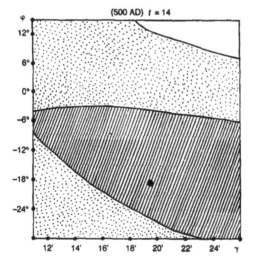

Figure 7.19.

Figures 7.6 through 7.23. The minimax deviation $\Delta(t, \gamma, \varphi)$ dependence on the a priori date t and the parameters of the systematic error γ and φ. At fixed t the rectangular confidence neighborhood of the point $(\gamma_{stat}^{Zod}\,^A(t), \varphi_{stat}^{Zod}\,^A(t))$ (marked bold) in the plane with the coordinates (γ, φ) is shown. The crosshatched domains are the sets of the parameters that lead to the minimax latitudinal deviation below 10′. The domains where the deviation is below 15′ are hatched, and those with the deviation below 20′ are dotted. The values of t for which the crosshatched domains are not void constitute the interval of admissible dates for the catalog. The figures cover the intervals 100–1800 AD.

4. Analysis of stability of the statistical dating procedure

1. Since some parameters that determined processing data in the above procedure were chosen quite arbitrarily, and some others were found from statistical methods, we need to check stability of the ensuing dating interval with respect to variations of these parameters.

2. The confidence level ε is generally chosen arbitrarily. Recall that in statistics the sense of ε is the probability of a mistake; thus, $\varepsilon = 0.1$ means that we admit a mistake with the probability 0.1. The less is ε, the wider is the confidence interval. The dependence of the length of the confidence interval on ε was studied in Chapters 5 and 6, see, in particular, Table 6.3.

We have already noted that *the dating interval is the same at all $\varepsilon \leq 0.1$*; this statement follows from the position of the intervals $S_t(10')$; see Figure 7.24.

What if we choose another level of claimed accuracy α? Let us take $\alpha = 15'$ (the blank domain in Figure 7.24). Of course, the interval of admissible dates expands; again, the upper bound of the interval does not depend on ε and is equal to 3 (1600 AD). The lower bound depends but slightly on ε: it is equal to 16.3 (270 AD) at $\varepsilon = 0.1$, and to 16.5 (250 AD) at $\varepsilon = 0.005$.

Thus, the choice of ε practically does not affect the ensuing dating interval. We have also shown how variation of the "claimed accuracy" α affects the dating interval; namely, if we raise α from $10'$ to $15'$ *the ensuing dating interval does not cover the traditional epoch of Ptolemy, not to mention of Hipparchus.*

3. Another to some extent arbitrary choice is the one of the informative kernel. Indeed, we have excluded four named stars, Canopus, Vindemiatrix, Sirius and Aquila; the first two were excluded for reasons "extrinsic" to our investigation, but Sirius and Aquila were excluded because the group errors for their vicinities did not coincide with the one appropriate to *Zod A*. We have shown in Chapter 6 that there are at least two more stars, Lyra and Capella, in whose vicinities the group errors may differ from the one in *Zod A*; the modality of this statement is due to the fact that we do not know the real values of these errors. Furthermore, these two stars are far from the zodiac and are close to the poorly measured domain M.

So let us look at what dating interval we get if we exclude these two stars, thus leaving six stars in the informative kernel (Arcturus, Regulus, Antares, Spica, Aselli and Procyon). The results of application of the dating procedure in this case are represented in Figure 7.25, in the fashion similar to Figure 7.24. Although the domains where the maximum latitudinal deviations do not exceed $10'$ and $15'$ expand, the bounds of the interval of admissible dates do not move much. Namely, the upper bounds remain unaltered (both for $10'$ and $15'$ levels), the lower bound of the interval for $\alpha = 15'$ also remains the same, and the lower bound for $\alpha = 10'$ moves to the past by at most 100 years (the particular variation depends on the confidence level $1 - \varepsilon$).

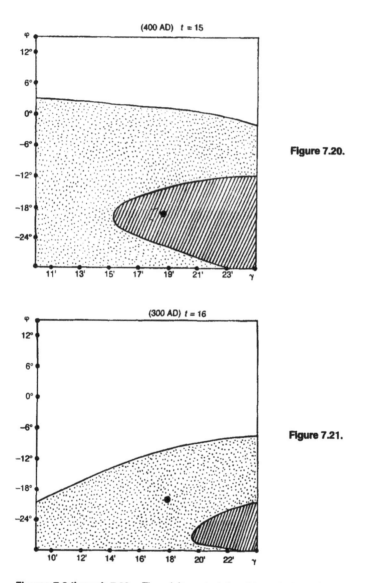

Figure 7.20.

Figure 7.21.

Figures 7.6 through 7.23. The minimax deviation $\Delta(t, \gamma, \varphi)$ dependence on the a priori date t and the parameters of the systematic error γ and φ. At fixed t the rectangular confidence neighborhood of the point $(\gamma_{stat}^{Zod\ A}(t), \varphi_{stat}^{Zod\ A}(t))$ (marked bold) in the plane with the coordinates (γ, φ) is shown. The crosshatched domains are the sets of the parameters that lead to the minimax latitudinal deviation below 10′. The domains where the deviation is below 15′ are hatched, and those with the deviation below 20′ are dotted. The values of t for which the crosshatched domains are not void constitute the interval of admissible dates for the catalog. The figures cover the intervals 100–1800 AD.

Thus, if we only take into account the six named stars, we must conclude that the catalog of the *Almagest* could not have been compiled before 500 AD.

4. We must also consider the possibility that the proposed dating interval is in fact generated by the motion of a single star; in this case any distortion in the coordinates of such a star could distort gravely the dating interval. The only candidate for such domination is Arcturus, the fastest star in the informative kernel, whose vicinity, by the way, is not very well measured (see Chapter 6). Let us find the dating interval that ensues after exclusion of Arcturus from the informative kernel. Of course, the length of the dating interval must increase (roughly speaking, it is inversely proportional to the maximum velocity of stars in the informative kernel). Figure 7.26 displays the ensuing picture. It is obvious from this figure that even after the exclusion of Arcturus the lower bound of the dating interval for $\alpha = 10'$ is not below 300 AD ($t = 16$) if we admit $\varepsilon \geq 0.05$, and only at $\varepsilon < 0.01$ does the interval capture 200 AD. Thus, even in the worst case, the interval does not capture Ptolemy's epoch.

At $\alpha = 15'$, the dating interval reaches 100 BC ($t = 20$) at $\varepsilon \geq 0.05$ and 200 BC at $\varepsilon \leq 0.01$. In this case, Hipparchus' epoch is out of possible dates. It seems only relevant to make the following remark: the level $\varepsilon = 0.05$ represents a fairly high precision for historical investigations; in fact, this confidence level is characteristic for technical applications. For comparison, the confidence level chosen in Ref. 27 is 0.8 ($\varepsilon = 0.2$). Thus, the above conclusions are highly plausible.

Thus, neither variations of the confidence level, nor of the structure of the informative kernel or of the claimed accuracy of the catalog alter our main conclusion: *the catalog of the Almagest was compiled much later than the traditional epoch of Ptolemy (the 1st–2nd centuries AD).*

5. Geometric dating procedure

The conclusions drawn in the previous sections are of statistical nature. Namely, both the values of the group errors are determined with statistical errors, and the conclusions concerning the coincidence of group errors in various constellations may be, in principle, false, though with small probability. We have analyzed in Section 4 the stability of previously obtained results; however, in order to avoid statistically possible errors, in this section we abandon statistics completely, and confine ourselves to purely geometric constructions.

Let us consider the "minimax latitudinal residual" of the above informative kernel consisting of the eight named stars,

(1) $$\delta(t) = \min \Delta(t, \gamma, \varphi)$$

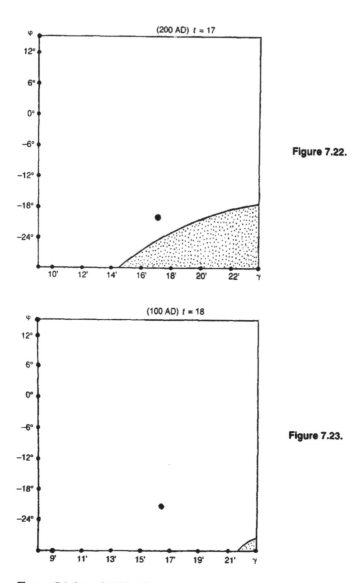

Figure 7.22.

Figure 7.23.

Figures 7.6 through 7.23. The minimax deviation $\Delta(t, \gamma, \varphi)$ dependence on the a priori date t and the parameters of the systematic error γ and φ. At fixed t the rectangular confidence neighborhood of the point $(\gamma_{stat}^{Zod} {}^A(t), \varphi_{stat}^{Zod} {}^A(t))$ (marked bold) in the plane with the coordinates (γ, φ) is shown. The crosshatched domains are the sets of the parameters that lead to the minimax latitudinal deviation below 10′. The domains where the deviation is below 15′ are hatched, and those with the deviation below 20′ are dotted. The values of t for which the crosshatched domains are not void constitute the interval of admissible dates for the catalog. The figures cover the intervals 100–1800 AD.

where the minimum is taken over all possible values of γ and φ. Let us compare this definition with (3.1). The only difference is in the range of the parameter γ: in (3.1) γ ranges over the confidence interval containing $\gamma_{stat}(t)$, while in (1) there is no such limitation. Hence,

$$(2) \qquad\qquad\qquad \delta(t) \leq \Delta(t)$$

Denote by $\gamma_{geom}(t)$ and $\varphi_{geom}(t)$ the values of the parameters that provide the minimum value in the right side of (1). A possible ambiguity in the definition of $\gamma_{geom}(t)$ and $\varphi_{geom}(t)$ will lead to no confusion. The situation here is much similar to the one we encountered in Section 3. There, lifting the limitations on the parameter φ and leaving the limitations on γ fixed, we obtained a fairly long dating interval, the length being in no way affected by the previously calculated statistical characteristics. We may regard $\gamma_{geom}(t)$ and $\varphi_{geom}(t)$, say, as the parameters that determine the conditional group error for the informative kernel (with the condition that the catalog was compiled at the moment t). It is natural to treat the set of all moments t with $\delta(t) \leq 10'$ as the dating interval. To find this interval, we depict in Figures 7.27–7.29 the $\delta(t)$ dependence, for which we used (1) (the values of $\Delta(t, \gamma, \varphi)$ had been found from (3.1), and the minima were found by the exhaustive search through the values of γ and φ), and the $\gamma_{geom}(t)$ and $\varphi_{geom}(t)$ dependencies. For comparison, we depict in Figure 7.28 the function $\gamma_{stat}(t)$ and the confidence zone (see Section 4), as well as the set of all pairs (t, γ) such that $\Delta(t, \gamma, \varphi) < 10'$ for some φ. The graphs show that *the geometric dating procedure does not expand the interval of admissible dates.* This, in particular, corroborates the statement that the parameters $\gamma_{stat}^{Zod\ A}$ we have found from statistics do match the group error for the collection of the named stars as well. Moreover, this proves that outside the temporal interval 600 AD–1300 AD the true positions of stars never match the ones given in the *Almagest* so that all stars of the informative kernel had at most 10' latitudinal residuals.

In conclusion, we give also the dependence on t, the a priori date, of the individual latitudinal residuals for the eight stars of the informative kernel at fixed $\gamma = 20'$ and $\varphi = 0$ (Figure 7.30). The upper envelope of the graphs is similar to the curve depicting the dependence of the minimum residual on the a priori date t (Figure 7.25) for the most part of the temporal interval after 1 AD. This is connected with the fact that the value $\gamma = 20'$ is close to $\gamma_{geom}(t)$ and $\varphi = 0$ is sufficiently close to $\varphi_{geom}(t)$ in a large part of this interval (note that the picture is but slightly sensitive to variations of φ). Figure 7.30 shows for which particular stars of the informative kernel the latitudinal deviation $\delta(t)$ attains the minimax value for various a priori dates t. Note the concentration of zero latitudinal deviations about $t = 10$ (900 AD) in Figure 7.30. This value of the a priori date yields almost zero deviations of three stars of the informative kernels at once: Arcturus (α Boo), Regulus (α Leo) and Procyon

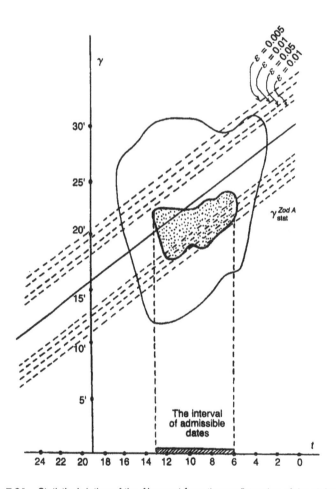

Figure 7.24. Statistical dating of the *Almagest* from the configuration of the eight named stars. Dashed lines are boundaries of confidence intervals about the optimal estimate $\gamma_{stat}^{Zod\ A}$. The dotted domains correspond to minimax latitudinal deviations below 10′, and the domains bounded by the solid line to the deviation below 15′. The projection of the intersection of the dotted domain with the confidence zone on the time axis is the interval of admissible dates, 600–1300 AD. It is obvious that the interval is irrelevant to the choice of the confidence level.

(α CMi). As for the rest of the stars, the latitudinal residuals only vanishes for Aselli (γ Can) near 1 AD.

It is interesting to compare the concentration of zero deviations with the fact that Arcturus and Regulus (together with Sirius) were distinguished in antique astronomy. Thus, Arcturus, the brightest star of the Northern hemisphere, was, apparently, the first star that received a name in antique astronomy (it was mentioned in the very first poetic description of the starry

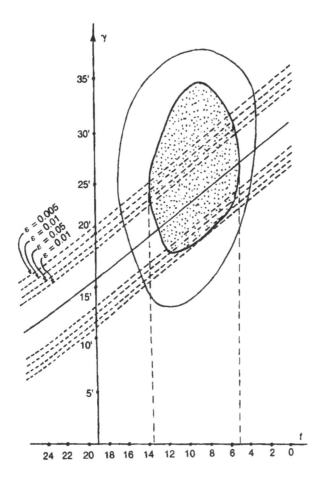

Figure 7.25. Statistical date for the *Almagest* from the collection of six named stars (Lyra and Capella excluded). Notation is as in Figure 7.24. The interval of admissible dates is practically the same.

sky, the poem of Aratus). Regulus is the star that was used as the starting point for measurement of coordinates of the rest of the stars and planets. A special section of the *Almagest* is devoted to measuring the position of Regulus.

6. Stability of the geometric dating procedure. Influence of possible instrumental errors on the results of dating

1. The confidence level never appears in the geometric dating procedure, so we are left to check the stability with respect to the accuracy and to the structure of the informative kernel. The conclusions here are very similar to

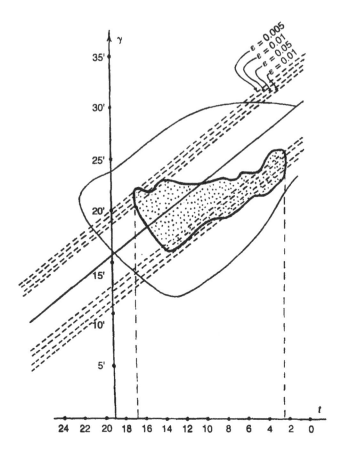

Figure 7.26. Statistical date for the *Almagest* from the collection of seven named stars (Arcturus excluded). Notation is as in Figures 7.24 and 7.25. Naturally, the dating interval expands; however, it does not cover 200 AD (not to mention 137 AD, the traditional date for compilation of the *Almagest*).

that of Section 4. Thus, increasing the accuracy level from 10′ to 15′ moves the lower bound of the dating interval to 250 AD (like in Section 4). For the informative kernel consisting of the six stars contained in *Zod A* or near, the dating interval is wider by approximately 100 years; excluding Arcturus from the informative kernel leads to the dating interval 200 AD–1600 AD.

Thus, in neither case does the dating interval that ensues from the geometric dating procedure cover the epoch of Ptolemy, not to mention Hipparchus.

In addition to these results on stability, that strengthen the results of Section 4 (because the "geometric" interval cannot be narrower than the "statisti-

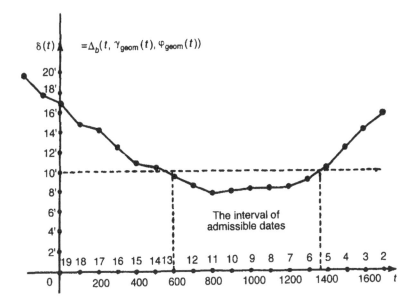

Figure 7.27. Geometric dating procedure. Minimax latitudinal deviation dependence on the a priori date. The interval where the graph lies below the 10′ level is the interval of admissible dates for the *Almagest*, 600–1300 AD. The interval coincides with the one obtained from the statistical dating procedure.

cal"), let us prove the stability of the geometric dating procedure with respect to possible instrumental errors.

The above dating method is based on the account of observer's error in determination of the position of the pole of the ecliptic, found from investigation of all possible turns of the celestial sphere (in different terms, of all rigid orthogonal turns of the coordinate frame). If we find a turn (which we describe in terms of the vector of displacement of the pole, with the coordinates γ and φ) that reduces the maximum latitudinal deviation (over the informative kernel or over zodiacal stars, etc.) to below the level Δ ($\Delta = 10′$ for the *Almagest*), then we carry out compensation for this turn and use this improved data for dating the catalog and drawing other conclusions.

In all the above cases, rigid turns of the celestial sphere sufficed to reduce the maximum latitudinal deviation to the level Δ, thus enabling us to confirm the claimed accuracy of the catalog and to apply the dating procedure. Meanwhile, the observer could use an imperfect instrument (say, an astrolabe), with metallic rings that are not quite round, say, oblate. Furthermore, some planes determined by the instrument and supposed to be perpendicular may be not quite so; also, slightly different scales could be plotted along dif-

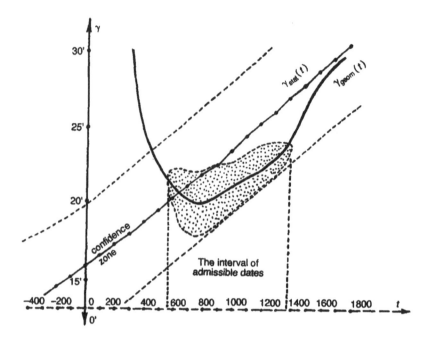

Figure 7.28. The optimal statistical estimate $\gamma_{stat}(t)$ and the optimal geometric value $\gamma_{geom}(t)$ dependencies on the a priori date. The dashed lines bound the confidence zone; the dotted domain corresponds to the latitudinal deviation below 10′.

ferent axes. Thus, the instrument (and the coordinate frame it determines) could be subject to a transformation, which could distort the results of measurements. The natural question arises: How could minor deformations of the instrument (and of the frame it determines) affect the results of measurements? How large must the deformations be to effect substantially the results of measurements? In this section we give complete answers to these questions.

2. Let us formulate the problem in mathematical terms. Suppose we are given a sphere in three-dimensional Euclidean space endowed with a Cartesian coordinate system with the origin at the center of the sphere. The coordinate axes determine three pairwise orthogonal coordinate planes. The process of measurement of ecliptic coordinates of a star consists in projecting the star to the sphere from its center (Figure 7.31). The coordinates (say, spherical) of the projection A of the star on the sphere are treated as the coordinates of the star, in particular, are included in the catalog. Let us assume for simplicity that the axis Oz is directed towards the pole of the

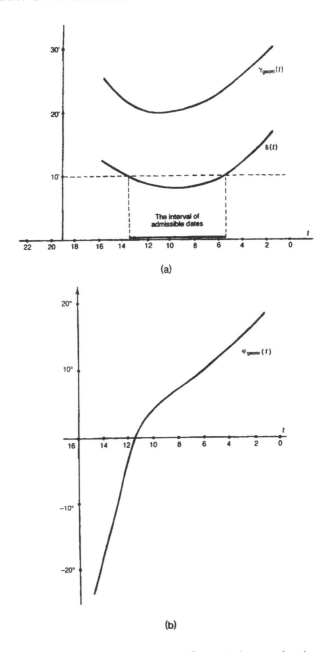

Figure 7.29. Geometric dating procedure. (a) The optimal geometric value γ and the corresponding minimum latitudinal deviation dependencies on the a priori date. (b) The optimal geometric value of φ dependence on the a priori date.

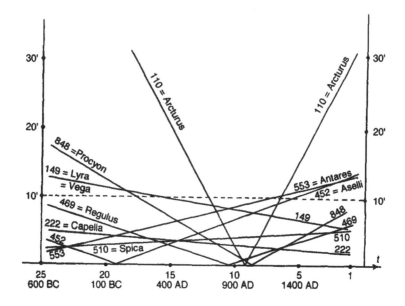

Figure 7.30. Individual latitudinal deviations of the eight named stars of the *Almagest* dependencies on the a priori date at *t* fixed parameters of the systematic error. $\beta \approx 0'$, $\gamma = 21'$.

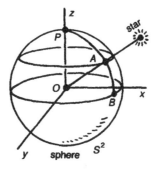

Figure 7.31. Measurement of ecliptic latitude of a star.

ecliptic P, and the coordinate plane Oxy bisects the sphere along the ecliptic. Since latitudes are more reliable data (see the discussion hereof above), we will only consider the latitude of the star A. The latitude is measured along the meridian through the pole of the ecliptic P and the point A; zero latitude corresponds to the ecliptic. In Figure 7.31, the ecliptic latitude of the star A is measured by the length of the arc AB.

Measurement of stellar coordinates implies the assumption that the observer's instrument determines an ideal spherical coordinate system in the three-dimensional space. However, the real instrument may be slightly deformed. Neglecting the second-order infinitesimal effects, we may assume that the deformation results in a linear transformation of the Euclidean coordinate system; it is natural to assume that the transformation is close to the identity transformation, because a too large deformation of the instrument would have been noted by the observer who claimed, say, the 10'-accuracy of measurement. Furthermore, even if the deformation of the coordinate system involves small nonlinear perturbations, we may consider the first approximation, that is, the linear approximation of the distortions.

A linear transformation of the three-dimensional space (fixing the origin) is determined by a 3×3-matrix

$$(1) \qquad C = \begin{pmatrix} c_{11} & c_{12} & c_{13} \\ c_{21} & c_{22} & c_{23} \\ c_{31} & c_{32} & c_{33} \end{pmatrix}$$

The transformation distorts the original Euclidean coordinate system; it follows from elementary theory of quadratic forms that a nondegenerate linear transformation takes the sphere to an ellipsoid (Figure 7.32). Thus, although the original coordinate axes are taken to some lines that need not be orthogonal, the three axes of the ellipsoid are orthogonal (the lines x', y', z' in Figure 7.32). Thus, we may assume that the transformation first turns the sphere, taking the axes x, y, z to the axes x', y', z' (an orthogonal transformation), and then dilates along the three axes with some coefficients λ_1, λ_2, λ_3. The latter transformation is determined by the diagonal matrix

$$(2) \qquad R = \begin{pmatrix} \lambda_1 & 0 & 0 \\ 0 & \lambda_2 & 0 \\ 0 & 0 & \lambda_3 \end{pmatrix} = \mathrm{diag}(\lambda_1, \lambda_2, \lambda_3)$$

The dilation coefficients λ_1, λ_2, λ_3 are some real numbers; from the sense of our problem, they are all positive.

3. We have already investigated in the previous sections the deformations due to turns of the coordinate frame, so now we may focus our attention at the second transformation, the dilation determined by the matrix R.

Thus, we may assume without loss of generality that the deformation of the instrument, inducing the linear transformation of the three-dimensional Euclidean coordinate frame in space, is determined by the dilation R with coefficients λ_1, λ_2, λ_3. Note that the three coefficients may be more than 1,

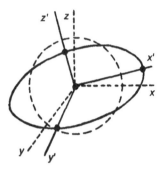

Figure 7.32. Variation of the coordinate frame due to a small deformation of the astronomic instrument.

equal to 1, or less than 1 (independently of each other), so as we speak of dilation, we do not necessarily mean an actual dilation (increase of linear size along an axis); if a coefficient is less than 1, we in fact deal with a contraction along the corresponding axis.

4. Let us now discuss in more details the measurement of coordinates of a star in the distorted coordinate system, which we will call *ellipsoidal*. Consider the plane through the center O, the star A and the pole of the ecliptic P. The plane bisects the ellipsoid along an ellipse, depicted in Figure 7.33 by a solid line. The section of the celestial sphere (which we in fact cannot treat as an ideal sphere corresponding to an ideal, not deformed instrument) is depicted in Figure 7.33 by the dashed circumference. We are only interested in latitudes, so recall that the latitudes are counted from the ecliptic, that is, from the point M in Figure 7.33. The observer divided the arc MP' into 90 equal parts, thus graduating the ring (the ellipse). Since he graduated the ellipse, but not a circumference, the uniform (in length) divisions represent slightly distorted angles; the resulting angular division is not uniform (which is left unnoticed by the observer, for otherwise he would have corrected the division).

Observing the star A, the observer marks the position A' in his ellipsoidal instrument and measures the arc $A'M$, thus obtaining the latitude of the star (true as he thinks), and writes it down in the catalog; of course, the catalog is compiled under the assumption that the instrument is ideal, so the coordinate written therein is supposed to be plotted along the ideal circumference (in Figure 7.33, the arc $A''M$). Thus, the observer displaces the position of the star. The ensuing transformation of the circumference $A \to A''$ is, of course, not linear. It may be extended to a transformation of the whole space onto itself (fixing the origin). Since we have fixed that the deformation of the instrument is small, we may confine ourselves to linear approximation, and consider

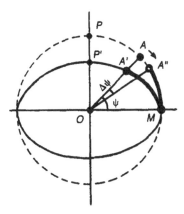

Figure 7.33. Deformation ellipsoid.

the linear component of the transformation instead of the transformation it-self. This linear part is the dilation along the three pairwise orthogonal axes with the coefficients λ_1, λ_2, λ_3. Thus, we come back to the statement of the problem as in Subsections 2 and 3 of this section (actually, we have also computed the exact distortions introduced by the nonlinear transformation, though we do not adduce the results here).

5. Thus, we consider the linear transformation of the three-dimensional space determined by three coefficients λ_1, λ_2, λ_3, that is, by the matrix $R = \mathrm{diag}(\lambda_1, \lambda_2, \lambda_3)$. How can we estimate the ensuing distortion of the angles? Let ψ be the true value of latitude of a star and ψ' be the value obtained from measurement with the help of the deformed instrument; the difference $\Delta\psi = \psi' - \psi$ is an estimate for the distortion. From the geometric point of view, the distortion of the angle is determined by the angle $\Delta\psi$ between the direction towards the true position of the star and the directions that reads from the instrument.

Further, we will consider the ellipse in the two-dimensional plane (Figure 7.34). Cancelling the previous notation, let us introduce in the plane Cartesian coordinates x, y and consider the linear transformation $R = \begin{pmatrix} \lambda_1 & 0 \\ 0 & \lambda_2 \end{pmatrix}$ determined by the coefficients λ_1 and λ_2 of dilation along the axes x and y. Let $\mathbf{a} = (x, y)$ be the radius vector of the point in the unit cir-cumference that determines the true position of a star A, and $\mathbf{b} = (\lambda_1 x, \lambda_2 y)$ be the radius vector that corresponds to the distorted position. Our aim now is to calculate the angle $\Delta\psi$ as a function of the true latitude ψ and the coefficients λ_1 and λ_2.

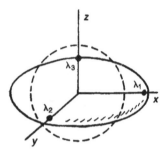

Figure 7.34. Distortion of stellar coordinates due to a small deformation of an astronomic instrument.

6.   By an elementary theorem of analytic geometry, $\cos \Delta \psi$ is equal to the scalar product $(\mathbf{a}, \mathbf{b})$ divided by the length of $\mathbf{b}$ (of course, we assume the radius of the circumference OM equal to 1, which we always may obtain by an appropriate choice of scaling). Thus,

$$(3) \qquad \cos \Delta \psi = \frac{\lambda_1 x^2 + \lambda_2 y^2}{\sqrt{\lambda_1^2 x^2 + \lambda_2^2 y^2}}$$

Put $\lambda = \lambda_1/\lambda_2$ and $\varepsilon = \lambda - 1$. Then

$$(4) \quad \cos \Delta \psi = \frac{\lambda x^2 + y^2}{\sqrt{\lambda^2 x^2 + y^2}} = \frac{\lambda x^2 + y^2 + \varepsilon y^2}{\sqrt{x^2 + y^2 + 2\varepsilon y^2 + y^2}} = \frac{1 + \varepsilon y^2}{\sqrt{1 + 2\varepsilon y^2 + \varepsilon^2 y^2}}$$

Put $m = 1/\cos \Delta \psi$; clearly, $m \geq 1$. Taking the squares of both sides, we get

$$(5) \qquad 1 + 2\varepsilon y^2 + \varepsilon^2 y^2 = m^2 + 2m^2 \varepsilon y^2 + m^2 \varepsilon^2 y^4$$

Hence

$$(6) \qquad \varepsilon = \frac{m^2 - 1}{1 - m^2 y^2} + \sqrt{\frac{m^2 - 1}{(1 - m^2 y^2) y^2} + \left(\frac{m^2 - 1}{1 - m^2 y^2}\right)^2}$$

Since we assume that $\Delta \psi$ is small,

$$(7) \qquad m \approx 1 + \frac{(\Delta \psi)^2}{2}$$

so

(8)
$$m - 1 \approx \frac{(\Delta\psi)^2}{2}$$

and

(9)
$$1 - m^2 y^2 = 1 - y^2$$

Finally, at small $\Delta\psi$ we have

(10)
$$\varepsilon \approx \sqrt{\frac{m^2 - 1}{(1 - m^2 y^2) y^2}} \approx \sqrt{\frac{(m - 1)(m + 1)}{(1 - m^2 y^2) y^2}} \approx \sqrt{\frac{(\Delta\psi)^2}{(1 - y^2) y^2}} \approx \frac{\Delta\psi}{y\sqrt{1 - y^2}}$$

But $y = \sin\psi$ (Figure 7.34), so $\sqrt{1 - y^2} = \cos\psi$. Thus, at small $\Delta\psi$ we get

(11)
$$\varepsilon \approx \frac{\Delta\psi}{\sin\psi \cos\psi} = \frac{2\Delta\psi}{\sin 2\psi}$$

Now let us estimate concrete values of ε. Recall that $\lambda_1/\lambda_2 = 1 + \varepsilon$, so ε indicates the degree of distortion of the coordinate system. It is convenient to use the values of angles in radians; we have $1° = \pi/180$ and $1' = 1°/60 \approx 3.14/(60 \cdot 180) \approx 2.9 \cdot 10^{-4}$, so $1' \approx 0.00029$.

The above expression for ε shows that ε increases as the star approaches the zodiac (the ecliptic) or the pole of the ecliptic, because in both cases $\sin 2\psi$ approaches zero. Hence, if ε is "reasonably small" (that is, if the deformations of the instrument are not visible by an unaided eye), then the latitudes of stars near the ecliptic and near the pole are distorted but little. The distortions are maximal for the stars remote both from the pole and the ecliptic. Now we will give exact quantitative estimates from the concrete contents of the star catalog of the *Almagest*. As is obvious from Figure 7.22, the maximum latitudinal deviation over the informative kernel of the *Almagest* increases fast to the left and to the right from the interval 700 AD–1300 AD. It is natural to ask whether it is possible to "suppress" this latitudinal deviation, say, near 1 AD, where the *Almagest* is traditionally attributed to. In other words, is it possible that a) the star catalog of the *Almagest* was compiled about 1 AD, but b) the observer used a deformed instrument, which introduced errors into the latitudes of stars; now, if we take the error into account, then will we find the date for the catalog about 1 AD?

So, let us assume that the instrument was deformed and try to compensate the latitudinal deviation that we have near 1 AD. It is rather large, not less than 35'. What ε must we admit to "suppress" it?

As we have already noted, it is hardly possible to "suppress" the latitudinal deviation for the stars whose latitudes are near 0° or 90°, so we should try the stars with latitudes about 30°–40°. The informative kernel contains Arcturus, whose latitude is 31°; moreover, due to its large velocity of proper motion, Arcturus contributes much to the maximum latitudinal deviation near 1 AD. As Figure 7.30 shows, the individual latitudinal deviation of Arcturus near 1 AD is just about 35'. So, let us find the value of ε that could bring about this deviation. The claimed accuracy of the catalog Δ is 10' (see above), so we need to lessen the latitude by about 25'. So, we need to choose ε so that $\Delta\psi$ should be equal to 25' (in radians, $\Delta\psi = 0.01$). Using (11), we get

$$(12) \qquad\qquad \varepsilon \approx 0.01/(\sin 30° \cos 30°) \approx 0.016$$

Thus, if we want the latitudinal deviation about 1 AD to be explained by a deformation of the observational instrument, we must admit the value for ε about 0.016. But this value for ε is too large! The fact is that if the radius of the astrolabe was about 50 cm, then the deformation had to be as large as to make the radius be about 51 cm. So, the error in the radius had to amount to 1 cm! It appears absolutely impossible to admit this error for an astronomic instrument; otherwise we will have to admit that in the times of Ptolemy cartwheels were manufactured with higher accuracy than astrolabe rings.

7.   Our conclusion is that no reasonable instrumental errors may explain the latitudinal deviation that arises near 1 AD for the informative kernel of the *Almagest*. So, the above results (including the dating interval for the *Almagest*, 700 AD–1300 AD) are stable with respect to reasonably admissible deformations of the observational instrument. In particular, no reasonable assumption about deformation of the instrument may lead to the date for the catalog about 1 AD.

7. Behavior of longitudes

We based our dating method on an analysis of ecliptic latitudes alone, the reasons for which we have been explained above. Nonetheless, we have carried out similar calculation (not much, in fact) for the dynamics of longitudes. The calculation showed that no refinement for the date of compilation of the *Almagest* within the interval 100 BC–1900 AD can be obtained from longitudes.

Denote by $L_i(t, \gamma, \varphi)$ the value of the longitude of the ith star after the turn of the celestial sphere through the angles γ and φ, that is, after compensation for the systematic error determined by the parameters γ and φ. In order to improve the accuracy of our conclusions, we consider six named stars that lie in the domain *Zod A* and near: Arcturus, Regulus, Antares, Spica, Aselli and Procyon. We have seen that for these stars, the group error is reliably equal to $\gamma_{\text{stat}}^{Zod\ A}$. Let us find $L_i(t, \gamma_{\text{stat}}^{Zod\ A}, \varphi_{\text{stat}}^{Zod\ A})$ for these stars, treating these as the latitudes of the stars at the moment t after compensation for the systematic error. Of course, in doing so we leave room for an error, and not a small error. There are at least two possible reasons for that. First, the parameter φ affects strongly the values of longitudes; meanwhile, as we have seen, this parameter is not determined stably, so we cannot be sure that the same value of this parameter is appropriate to all the six stars. Second, we did not analyze group longitudinal errors, which also can exist; see Ref. 22. An investigation of such errors requires introducing one more parameter to determine the group error. We can use as such the parameter τ (see Chapter 3), the angle of turn of the celestial sphere about the new poles of the ecliptic determined by the parameters γ and φ.

Put

$$(1) \qquad \Delta L_i(t) = L_i(t, \gamma_{\text{stat}}^{Zod\ A}(t), \varphi_{\text{stat}}^{Zod\ A}(t)) - l_i$$

We may represent $\Delta L_i(t)$ as the sum of an "almost linear" function (variation of longitudes due to precession) and an irregular addend due to various errors. In order to eliminate the effect of precession and of the possible systematic error τ, let us define

$$(2) \qquad \overline{\Delta L}(t) = \frac{1}{6} \sum_{i=1}^{6} \Delta L_i(t)$$

This quantity reflects practically precisely the effect of precession; put also

$$(3) \qquad \Delta L_i^0(t) = \Delta L_i(t) - \overline{\Delta L}(t)$$

Figure 7.35 exhibits variation of $\Delta L_i(t)$ for each of the six stars (Baily's numbers are given in the figure: 110 for Arcturus, 469 for Regulus, 848 for Procyon, 553 for Antares, 510 for Spica, 452 for Aselli). It is obvious from the figure that the functions $\Delta L_i(t)$ vary very slowly with time. It turns out that after compensation for precession, the fast stars become slow (in longitudes); for example, the compensated longitudinal velocities of Arcturus and Regulus are almost equal. The fastest star is now Procyon, whose longitude varies by $17'$ in 3000 years (from 1100 BC to 1900 AD). Clearly, this sluggish variation is hardly useful for dating purposes. Figure 7.36 exposes two graphs that, in principle, could be useful for dating, but the form of which shows that they are not. Namely, we consider two functions,

$$(4) \qquad \Delta L_{\text{max}}(t) = \max_i |\Delta L_i^0(t)|$$

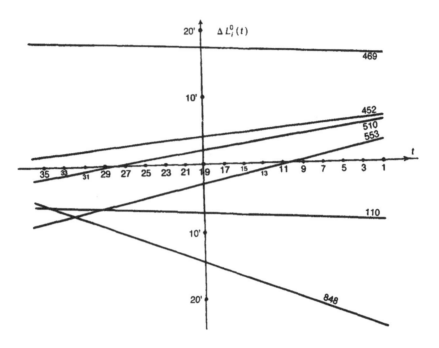

Figure 7.35. Behavior of longitudes of six named stars, Arcturus, Regulus, Procyon, Antares, Spica, and Aselli.

the maximum (in absolute value) deviation of true longitudes from the ones given in the *Almagest* (after compensation for precession), and

(5) $$\Delta L^0(t) = \max_i \Delta L_i^0(t) - \min_i \Delta L_i^0(t)$$

the difference between the maximum and the minimum deviations. The function $\Delta L_{max}(t)$ attains its maximum value at $t = 15$ (400 AD), and $\Delta L^0(t)$ at $t = 32.5$ (2350 BC). Both functions are comparatively large: $\Delta L^0(t) \geq 25'$ (and starting with the epoch of Hipparchus, $\Delta L^0(t) \geq 30'$), and $\Delta L_{max}(t) \geq 17'$.

Thus, our investigation leads to the conclusion that using longitudes for dating the catalog is apparently senseless.

8. Behavior of angular deviations in the configuration formed by the informative kernel

We have discussed in Chapter 3 the possibility of dating the catalog from a comparative analysis of two stellar configurations, the one fixed in the

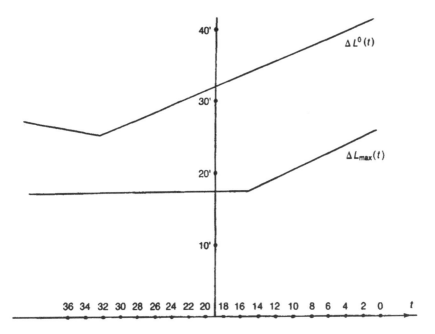

Figure 7.36. Two characteristics of longitudinal errors dependencies on the a priori dates. It follows that the collection of longitudes given in the *Almagest* is of little use for dating purposes.

Almagest and the moving true configuration. This comparison may be carried out without use of Newcomb's theory, just from differences of angular distances in the configurations. We have revealed the following difficulties that hinder application of this approach: possible mistakes in identification of stars, low accuracy of coordinates in the catalog that leads to too long dating intervals, and the impossibility of separating well-measured and poorly measured coordinates (say, latitudes and longitudes).

The choice of the configuration formed by the stars in the informative kernel of the catalog lifts the first two obstacles. Indeed, identification of named stars raises no doubts, and in accordance with our basic assumption, the accuracy of measurement of their coordinates must be sufficiently high (at least, in what concerns latitudes). Moreover, the informative kernel includes two fast stars, Arcturus and Procyon. Of course, the unknown accuracy of measurement of longitudes may bring about an inaccuracy in the date which we cannot estimate. Nevertheless, the absence of necessity to estimate group errors characteristic for this approach makes the results of its application interesting, although, as we have noted, we cannot estimate their accuracy. Therefore we only present the results of calculations for the configurations formed by the eight stars and by the six stars.

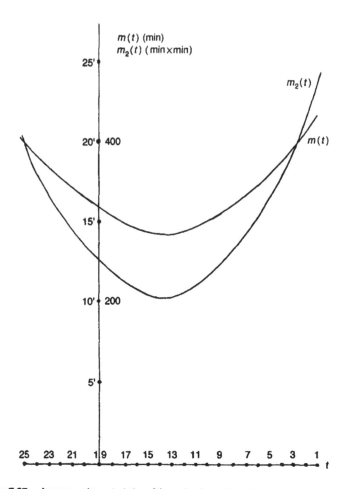

Figure 7.37. Accuracy characteristics of the collection of the eight named stars' dependencies on the a priori date. The graphs have minima at $t = 14$ (500 AD), and the corresponding confidence intervals cover all the historical period. The result is in no contradiction with the previously obtained dating interval for the *Almagest* (600–1300 AD).

Let l_{ij}^A be the angular distance between the ith and the jth stars of the *Almagest*, and l_{ij}^t the angular distance between the true positions of the stars in the year t, $t = 1, \ldots, 25$. Denote by n the number of stars in the configuration in question, and put

$$(1) \qquad m_2(t) = \frac{2}{n(n-1)} \sum (l_{ij}^t - l_{ij}^A)^2$$

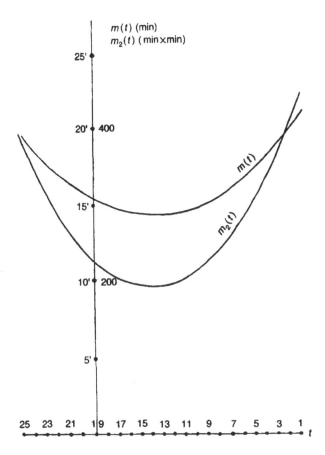

Figure 7.38. Accuracy characteristics of the collection of six named stars' (Lyra and Capella excluded) dependencies on the a priori date. The graphs have minima that are not clear cut and are about at $t = 15$ (400 AD). The corresponding confidence intervals cover all the historical period. The result is no contradiction with the previously obtained dating interval for the *Almagest*.

and

$$(2) \qquad m(t) = \sqrt{m_2(t)}$$

The quantity $m(t)$ may be treated as a distance between the true stellar configuration in the year t and the corresponding configuration in the *Almagest*. The point of minimum of the functions $m_2(t)$ and $m(t)$ should be close to the true date of compilation of the catalog. Figure 7.37 exposes the graphs of

$m_2(t)$ and $m(t)$ for the configuration of eight stars, and Figure 7.38 the graphs for the configuration of six stars.

In both cases the clearly cut minimum is at $t = 14$; here $m(t)$ is equal to 14′, which corresponds to the accuracy of 10′ in each coordinate. Clearly, this date is rather far from the traditional date of compilation of the *Almagest*. A certain displacement of this date to the past in comparison with the interval we have obtained from the analysis of latitudes may be explained by the fact (see Section 7 of this chapter) that the error in longitudes attains its minimum value about $t = 31$ (1200 BC). This date cannot be perspicuously explained, but since the minimum of longitudinal deviation is not very distinct (see Figures 7.35 and 7.36), the accuracy of these dates may amount to several thousands years. The minimum of latitudinal deviation falls on $t = 10$ (900 AD), and is much more clearly cut. As a result, the minimum of mean square angular distance falls to an intermediate point $t = 14$ (500 AD), much nearer to the clearly cut minimum for latitudes than to that for longitudes.

9. Brief conclusions

1. The dating interval that ensues from application of statistical and geometric dating procedures is 600–1300 AD.

2. Before 600 AD, the true positions of stars never match the positions given in the *Almagest* so that all stars in the informative kernel had latitudinal deviations not exceeding 10′.

3. The assumption that the true accuracy of the catalog of the *Almagest* is 15′ but not 10′ does not expand the interval of admissible dates so as to cover the epoch Ptolemy is traditionally attributed to (the 1st–2nd centuries AD).

4. Variations of the choice of named stars for the informative kernel also do not lead to an interval of admissible date covering the 1st or the 2nd centuries AD.

5. No compensation for reasonably admissible inaccuracies in manufacturing the observational instruments used for compilation of the catalog may expand or move the interval of admissible dates so as to cover the 1st or the 2nd century AD.

Chapter 8

Inclination of the Ecliptic in the *Almagest*

1. Inclination of the ecliptic in the *Almagest* and the systematic error

1.   The angle ε of inclination of the ecliptic to the equator is one of basic astronomic magnitudes. The knowledge of the value of this angle is necessary for determination of ecliptic coordinates of stars, whatever the method for finding these coordinates: the use of an astrolabe (as described in the *Almagest*), recalculation from equatorial coordinates with the help of a double-framed celestial globe (as was done in Middle Ages), or any other (see Introduction and Chapter 2).

In the text of the *Almagest*, the methods of measurement of the angle ε are discussed in details, and a description of the instruments used for this measurement is given (*Almagest*, Ref. 17, Chapter 1.12). It is stated that the measurement gives the value 11/83 of the full circle for 2ε, so in modern notation, $\varepsilon_A = 23°51'20''$; here ε_A is the value given in the *Almagest*.

As he compiled the catalog, the author of the *Almagest* had to use this value of ε, fixing it in his instrument (astrolabe, double-framed globe, etc.). An error in the value of ε introduces a turn of all the celestial sphere through the angle equal to the error. In other words, an inaccuracy in fixing the angle in the astronomic instrument brings about a systematic error in the coordinates of all stars in the catalog (more exactly, in the coordinates of the stars measured with the help of this instrument). It is easy to see that the error affects most the latitudes of stars. This error we have estimated in Chapter 6 as we found the values of $\gamma_{\text{stat}}(t)$ for various t and various regions of the celestial sphere. The dependence of the error on t is due to the fact that the true value of the angle ε varies with t; this dependence is monotone and is practically linear in the a priori dating interval $0 \leq t \leq 25$. The value fixed by the author of the catalog is either more or less than the true value, so the author, as he made the error, had either "rejuvenated" the catalog, or "made it older" with regard to the inclination of the ecliptic to the equator. Each of the possibilities could be realized with probability 1/2, and the first possibility did realize: the value of ε given in the *Almagest* is equal to the true value of $\varepsilon(t)$ for t approximately equal to 32 (1200 BC; see Chapter 6).

2. Suppose that the catalog of the *Almagest* was compiled in the year t, and its author considered the angle of inclination of the ecliptic to be equal to $23°51'20''$ (the value given in the *Almagest*) and tried to fix this angle in his astronomic instrument designed for determination (from an immediate observation or a recalculation) of ecliptic coordinates of stars. If we take into account a possible error $\pm\Delta\varepsilon$ in this fixing, determined by the accuracy of manufacturing the instrument, then the net error in fixing the inclination amounts to $\varepsilon_A - \varepsilon(t) \pm \Delta\varepsilon = 23°51'20'' - \varepsilon(t) \pm \Delta\varepsilon$. Let us compare this with the confidence zone $\gamma_{\text{stat}}(t) \pm \Delta\gamma$ for the systematic error γ found in Chapter 6 and with the set of all values of γ for which a matching configuration of the six stars of the informative kernel of the *Almagest* with the corresponding true configuration is possible within $10'$ latitudinal deviation (the latter set is empty unless $6 \leq t \leq 12$; see Chapter 7). As $\gamma_{\text{stat}}(t)$, we take the values determined for *Zod A*, because, as we have noted, the confidence zone for systematic error γ for this domain is narrower than in other parts of the catalog; furthermore, all stars of the informative kernel lie either in *Zod A* or immediately near it (see Chapter 7).

Figure 8.1 shows the confidence zone $\gamma_{\text{stat}}(t) \pm \Delta\gamma$ for the domain *Zod A* with the confidence level 0.002, the set of admissible values $\gamma_{\text{geom}}(t)$ from the geometrical dating procedure (that is, the set of all values for which the maximum latitudinal deviation of stars in the informative kernel does not exceed $10'$, see Chapter 7), and the graph of dependence of deviation of the value ε_A, for ε given in the *Almagest*, from the true value in the year t: $\gamma_{\text{Alm}}(t) = \varepsilon_A - \varepsilon(t)$. It is obvious from Figure 8.1 that the graph of $\gamma_{\text{Alm}}(t)$ is close to the "geometrically admissible" domain $(\gamma, t)_{\text{geom}}$ and to the confidence zone about $\gamma_{\text{stat}}(t)$, although it does not meet them. In order that they intersect, it is necessary to displace the graph $\gamma_{\text{Alm}}(t)$ by $2'5$ upwards; then the graph will intersect both the confidence zone and the "geometrically admissible" domain, which is nearer to the lower edge of the confidence zone (see Figure 8.1). After the displacement by $6'5$ upwards, the graph of $\gamma_{\text{Alm}}(t)$ practically coincides with the graph of $\gamma_{\text{stat}}(t)$ dependence, still intersecting with the "geometrically admissible" domain. The necessary displacement corresponds to the error $\Delta\varepsilon$ in fixing ε_A, and gives an idea of the accuracy with

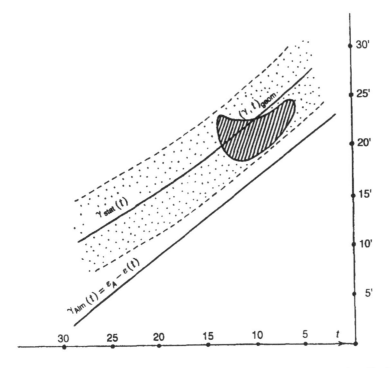

Figure 8.1. Confidence zone for $\gamma_{stat}(t)$ and the difference between the true value of inclination of the ecliptic and the value given in the *Almagest* dependence on the a priori date. The hatched domain correspond to the minimum latitudinal deviations below 10′.

Table 8.1.

Angle	Radius		
	50 cm	75 cm	1 m
360°	3 m 14 cm	4 m 71 cm	6 m 28 cm
5′30″	0.4 mm	0.5 mm	0.7 mm
10′	1.5 mm	2.2 mm	2.9 mm
1°	8.7 mm	13.0 mm	17.5 mm

which the instrument had been manufactured. Table 8.1 contains the lengths of arcs subtending the angles 2′5, 5′ and 10′ with the radius of the instrument equal to 50 cm, 75 cm, or 1 m.

It is obvious from Table 8.1 that the error $\Delta\varepsilon = 2'5 \div 5'$ in fixing the angle ε in the instrument is quite relevant not only for antiquity, but even for the Middle Ages; it corresponds to as small a linear error as $0.5 \div 1$ mm.

Thus, the values of inclination of the ecliptic that correspond to the previously determined group errors agree with the value of the angle given in the *Almagest*.

2. Zodiac in the *Almagest* and the Peters' sine curve

1. In the literature devoted to the *Almagest*, the so-called *Peters' sine curve* is known (see Ref. 22, p. 6), which implies the presence of certain systematic errors in the catalog of the *Almagest*. In this section we suggest an explanation for this sine curve.

2. Let Π be the position of the ecliptic at $t = 18$, that is, in 100 AD; mark the spring equinoctial point $Q(18)$ on Π.

Let us calculate for each of 350 zodiacal stars of the *Almagest* its latitudinal deviation and depict it in the graph; we plot the longitudes of stars along the horizontal axis and the latitudinal deviations along the vertical axis. As a result, we obtain a collection of points in the plane, which we call *the field of (latitudinal) errors*. Dividing the axis of longitudes into 10° long intervals and averaging in each interval, we construct the smoothing curve, shown in Figure 8.2. This curve may be approximated (according to minimum mean square deviation) by a sine curve, called *Peters' sine curve*.

A similar procedure may be applied to longitudinal deviations; the resulting curve is the dashed line in Figure 8.2. We will discuss this curve further. Figure 8.2 is taken from the Ref. 22.

Our aim is to give explanation for the form of these curves.

3. We begin with the *latitudinal Peters' curve*. There is a natural cause for systematic errors in latitudes of zodiacal stars, the error in the position of the ecliptic used by the observer in relation to the true position in the year of observations (which we a priori do not know).

Let us construct the above fields of latitudinal errors for each moment t_0 (Figure 8.3). The smoothing curve of averaged errors (the dashed line in Figure 8.3) we denote by $c(X, K(t_0, 0, 0))$; here X stands for the catalog of the *Almagest* and $K(t, \beta, \gamma)$ is the computed catalog of true positions of stars in the year t disturbed by the systematic error with the parameters β and γ (see Chapter 6). Thus, $K(t_0, 0, 0)$ is the catalog of true stellar positions in the year t_0.

As described in Chapter 6, in order to find the turn of the ecliptic bringing about the field of errors nearest to the given one (in the sense of mean square error), we need to solve a regression problem, using the two-parametric family of sine curves as an approximation. The first parameter is the amplitude of a sine curve, and the second is the phase. In Chapter 6, we have solved this problem for the catalog of the *Almagest* as a whole, as well as for various parts of the catalog, in particular, for the zodiacal domains, in which we

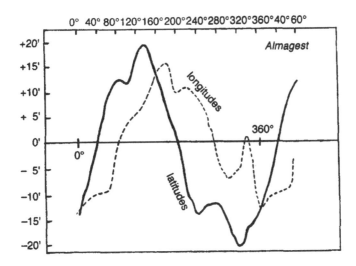

Figure 8.2.   Latitudinal and longitudinal smoothing Peters' curves for 100 AD.

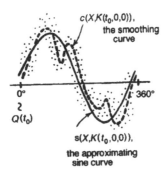

Figure 8.3.   Latitudinal and approximating sine curves for the field of latitudinal errors.

are interested at the moment. We denote the optimal approximating sine curve (solid line in Figure 8.3) by $s(X, K(t_0, 0, 0))$, and the parameters that determine it by A^* (the amplitude) and ϕ^* (the phase; see Figure 8.3).

4.   We should discuss the notion of the *phase* of the approximating sine curve. The fact is that the phase is only determined with accuracy within plus or minus 15° (at least). Figure 8.4 shows the true equator at the moment t_0, which (as was explained above) may be treated as coincident with the observer's equator, the true ecliptic in the year t, and the position of the ecliptic assumed by the observer. We know that the angle the observer's

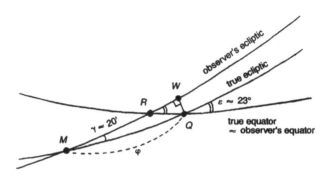

Figure 8.4. Geometric sense of the phase φ of the approximating sine curve.

ecliptic makes with the true ecliptic is about 20′ (the observer's error γ). The angle between the equator and the ecliptic is ε, about 23° ; it is not important, which of the two ecliptics is meant here, because the angle between them is small in relation to 23°. The arc in Figure 8.4 depicts the observer's error in the position of the spring equinoctial point; as we know, this error is about 10′. In this case, the arc distance WQ is about 10′ · sin 20°, that is, about 5′. Therefore, the arc distance ϕ (the arc MQ in Figure 8.4) is approximately 5′ · cot 20′, that is, about 15°. We are left to observe that the arc MQ is exactly the phase of the approximating sine curve (we count the phase of the sine curve from the spring equinoctial point $Q(t)$ in the true ecliptic $\Pi(t)$).

So, *several minutes' perturbations in determination of the ecliptic generate several degrees' perturbations of the phase of the approximating sine curve*; thus, the phase is unstable.

5. In the previous sections, we have estimated the interval of admissible dates for the catalog of the *Almagest*: t_0 is between 6 and 13 (600 AD–1300 AD). Therefore, a study of the approximating sine curves $s(X, K(t_0, 0, 0))$ is especially interesting for t_0 ranging in this interval. It turns out that the curves for t_0 in the interval vary but little; more precisely, the amplitudes A^* vary from 26′ (at $t_0 = 6$) to 20′ (at $t_0 = 13$), and the phase ϕ^* varies from −17° to −18° (counted from the appropriate equinoctial point $Q(t)$ in the ecliptic $\Pi(t)$). Therefore, we may choose as "typical" any smoothing curve $c(X, K(t_0, 0, 0))$ with $6 \leq t \leq 13$; it is natural to take as such the curve with $t_0 = 9$, the midpoint of the interval.

Let us look at the smoothing curve at $t_0 = 9$ before and after subtracting the approximating sine curve (that is, before and after compensation for the systematic error). The parameters of the optimal sine curve at $t_0 = 9$ are $A^* = 24′$ and $\phi^* = -17°$. The smoothing curve is depicted as a dashed line in Figure 8.5. Elimination of the observer's systematic error from the catalog

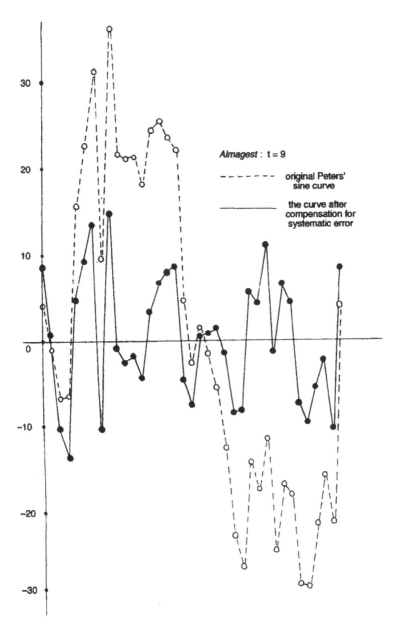

Figure 8.5. Collapse of Peters' sine curve after compensation for the systematic error.

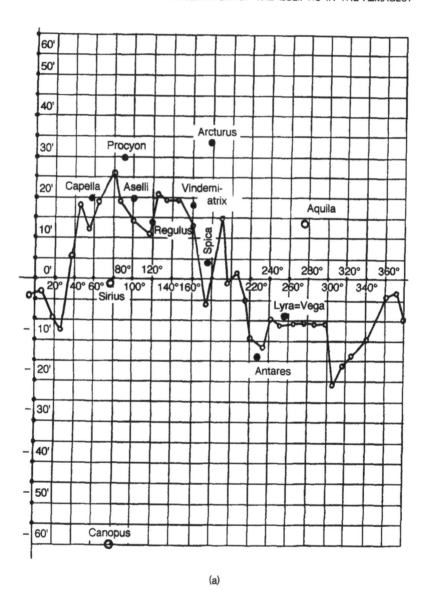

(a)

Figure 8.6. (a) Errors in latitudes of zodiacal stars dependencies on longitudes. The errors in latitudes of the twelve named stars are shown separately. This graph is similar to the one of Peters'. (b) Field of latitudinal errors for all stars of the *Almagest*, distributed in longitudes. Black dots denote zodiacal stars. The smoothing curve and the approximating sine curve are shown.

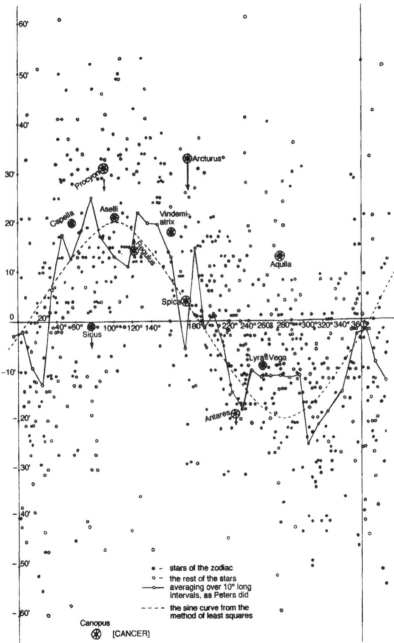

(b)

K is equivalent to subtraction of the optimal sine curve from the smoothing curve; as a result, the curve of latitudinal deviations assumes the form shown in Figure 8.5 (the solid line). *The difference between the dotted and the solid curves is obvious.* The latter oscillates about the axis of abscissas and corresponds to the zero mean observer's error in determination of the position of the ecliptic.

6.   Let us now return to the Peters' sine curve (latitudinal). Since Peters could omit some zodiacal stars in his calculations, we calculated anew the graph similar to that of Peters (Figure 8.2; here $t = 18$, which corresponds to 100 AD). In doing this, we took into account all zodiacal stars of the *Almagest* except several outlies and the stars with latitudinal deviation above $1°.5$. We used the data given in Ref. 22.

The result of our calculation is shown in Figure 8.6. The figure exhibits the field of latitudinal errors (at $t = 18$) for the zodiac of the *Almagest*.The field is represented by 350 points dispersed in the coordinate plane. The solid line is the smoothing curve $c(X, K(18, 0, 0))$. It is obvious that the curve is qualitatively similar to the Peters' curve in Figure 8.2; however, there are certain (not large) distinctions, apparently due to criteria of selection of zodiacal stars used by Peters, which we do not know.

The dashed line in Figure 8.6 represents the optimal approximating sine curve $s(X, K(18, 0, 0))$, with the amplitude $16'$ and the phase $-22°$ (cf. Chapter 6).

7.   Above, we studied the properties of the field of latitudinal errors in relation to the true year of observations t_0. Let us now consider the field of latitudinal errors for an arbitrary t. *The field of latitudinal errors in relation to the ecliptic $\Pi(t)$ is approximated by a sum of two sine curves.* The first sine curve is due to the observer's error in the year t_0; we have discussed it above. Its phase counted from the spring equinoctial point $Q(t)$ (in the ecliptic $\Pi(t)$) is the sum (approximately) of its phase in relation to the spring equinoctial point $Q(t_0)$ with the precession accumulated for the time $t - t_0$.

The other sine curve, s_{t, t_0}, is due to the deviation of the ecliptic $\Pi(t)$ from the ecliptic $\Pi(t_0)$; its amplitude is approximately equal to $47'' \cdot (t - t_0)$.

Thus, stating it a bit roughly, we may say that *the sine curve of Peters' type for the year t is approximately the sum of the Peters' type sine curve for the year t_0 and the sine curve due to the turn of the ecliptic for the time $t - t_0$ (in the interval of time from t to t_0).* This is a general statement, valid for all pairs of t and t_0.

8.   Let us now look at what approximating curve must result for $t = 18$ (100 AD). As we have seen, we should sum up two sine curves. The first of them corresponds to the true year of observations t_0, and the second to the year t for which we calculate the resulting approximating curve. We take as the "true year of observations" the value $t_0 = 9$ (approximately 1000 AD), the midpoint of the interval of admissible dates $6 \leq t \leq 13$ (600 AD–1300 AD) we

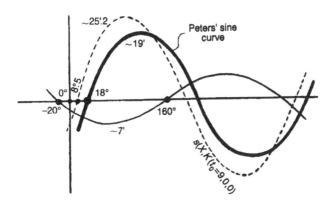

Figure 8.7. How the Peters' sine curve emerges: it is the sum of two sine curves.

have found above. The first sine curve (the dashed line in Figure 8.7) has the amplitude 24′ and the phase −5° (in relation to the spring equinoctial point of 100 AD). The second sine curve (thin solid line in Figure 8.7) corresponds to the choice $t = 18$ (100 AD, see above); its amplitude is approximately equal to 7′ ($\approx 47'' \cdot 9$), and the phase is approximately equal to 160° (see Chapter 1). Summing the two sine curves, we obtain the resulting approximating curve, shown in Figure 8.7 as a bold solid line—this is the Peters' sine curve.

9.   In conclusion, we turn to the "longitudinal Peters' curve" (dashed line in Figure 8.2). The above mechanism explains the "latitudinal sine curve", but, as is easy to see, it affects but little the longitudes of zodiacal stars, so the observer's error in determination of the position of the ecliptic does not imply the appearance of a notable longitudinal sine curve (although a curve with a very small amplitude may arise). As we have noted many times, the longitudes given in the *Almagest* are, apparently, not the original material, but the result of a recalculation (see, in particular, Ref. 1), so we based our study on the latitudes of the *Almagest* alone. Nonetheless, we will suggest a possible simple explanation for the longitudinal Peters' curve. Suppose that the observer had determined inaccurately the positions of the spring and fall equinoctial points, or, which is, in fact, the same, determined inaccurately the coordinates of the reference stars (note that the latitudes were counted from the ecliptic ring of the astronomic instrument, which was fixed with the never varying error, while the longitudes of stars were measured from various reference stars— otherwise the observer would have to measure angles that exceed 180°, which is extremely inconvenient—*Almagest*, Chapters VII.3, VII.4). An inaccuracy in measurement of the equinoctial points results in actual division of the ecliptic into two unequal parts by the points $Q(t_0)$ and $R(t_0)$ (Figure 8.8); here

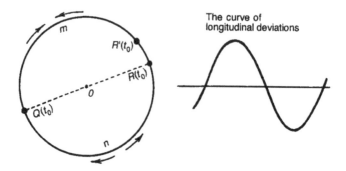

Figure 8.8. Explanation of longitudinal Peters' curve.

$R'(t_0)$ is the (erroneously) measured position of, say, the fall equinoctial point, and $R(t)$ is the true position. The length of the arc may not be large, about $10'-15'$, that is, within the accuracy of the *Almagest*; longitudes of some of the stars could be measured with reference to the spring equinoctial point Q (that is, with some reference stars), and the longitudes of the other with reference to the fall equinoctial point R' (that is, with reference to other reference stars). As a result, the measured longitudes of stars in the interval QmR' will be approximately $15'$ less, and in the interval QnR' will be approximately $15'$ more than the true ones. Therefore, plotting the longitudinal deviations of zodiacal stars, we obtain a sine-like curve (Figure 8.8). Note that the $10'-15'$ large error in longitudes is comparatively small, and this is the amplitude of the longitudinal Peters' curve in Figure 8.2.

Chapter **9**

Dating Other
Medieval Catalogs

1. Introduction

In the previous chapters we have described a method of statistical analysis for dating ancient star catalogs and applied it to dating the catalog of the *Almagest*. It appears to be of interest to apply this method to dating other star catalogs obtained with the help of astronomic instruments of the same type as the ones used by Ptolemy—that is, the catalogs compiled from observations by an unaided eye.

We applied our method to the catalogs of Ulugh Beg, Al Sûfi, Tycho Brahe, and Hevelius. The catalog of Al Sufi turned out to be a version of the catalog of the *Almagest* (this has been noted by some researchers, see Ref. 22, p. 7). Apparently, no detailed statistical analysis of coordinates (latitudes) of stars in the catalogs of Ulugh Beg, Tycho Brahe and Hevelius has been carried out before. The analysis showed that the accuracy of the catalogs is actually much lower than was accepted (see below); this difference is especially large for the catalog of Hevelius (100–200 times).

We found the dates for the catalogs of Tycho Brahe and Ulugh Beg. The date of observations of Tycho Brahe is known well, and the agreement of our date with this known date corroborates our method. In the case of Ulugh Beg, the interval of admissible dates we have obtained also covers the traditionally accepted date of compilation, 1437 AD. The interval overlaps strongly the interval of admissible dates for the catalog of the *Almagest* (see Section 3), and the two catalog have similar accuracies, so it may happen that the dates of compilation of the two catalogs are near.

2. Catalog of Tycho Brahe

1. General characteristics of the catalog and the results of dating.

We took for our study Kepler's 1628 edition of the catalog of Tycho Brahe reprinted in Ref. 28. The catalog is reduced for precession to 1600 AD. The structure of the catalog is similar to that of the catalog of the *Almagest*; even the enumeration of constellations is exactly as in the *Almagest* (except several constellations in the end of the catalog of the *Almagest*, absent in the catalog of Tycho Brahe). The total number of stars in the catalog is 1005. Basically, construction of the instruments used by Tycho Brahe is similar to that used in the times of Ptolemy, therefore, despite numerous improvements and high accuracy of manufacturing observational instruments, the accuracy of Brahe's measurements, 2' to 3', is comparable to the accuracy of the *Almagest* (10'–15'); a sharp improvement of accuracy followed due to the invention of telescopes.

It is known that Tycho Brahe carried out his observations in 1570–1600. As we date the catalog independently of the known dates, just on the basis of astronomic data given therein alone, we demonstrate the effectiveness of the suggested dating method in the problem with the known answer. The dating interval we obtained is 1510–1620; it is 110 years long and it covers the true date. Note that the length of the interval is six times less than the dating interval for the *Almagest* (approximately 700 years long), because the accuracy of observation here is 5 to 6 times better than that in the *Almagest*.

2. Analysis of latitudinal errors and deleting outlies.

For the same reasons as in the case of the *Almagest*, we only used latitudes of stars as we analyzed and dated the catalog of Tycho Brahe; we used the identifications of stars of the catalog given in Ref. 28.

There is a hypothesis that Brahe had in fact observed only 800 of 1005 stars included in the catalog (see Ref. 28, p. 126). In this case, the data contained in the catalog are not homogeneous. In order to educe the homogeneous part of the catalog, we constructed the histograms of frequencies of latitudinal errors separately for each of the domains A, Zod A, B, Zod B, C, D and M distinguished in Section 2.3. We calculated the ecliptic coordinates of stars for 1600 (that is, the catalog $K(t)$ for $t = 3$, see Section 1.5), and then compared the latitudes given in the catalog with the computed true latitudes. The histograms of frequencies of occurrence of errors are displayed in Figures 9.1–9.7; the scale of errors is plotted along the horizontal axis, and the value

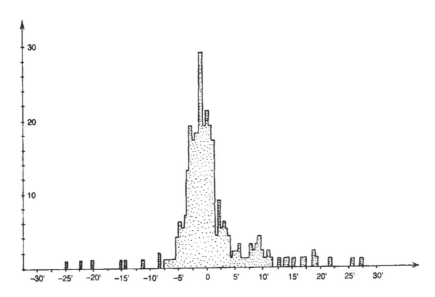

Figure 9.1. Histograms of frequencies of latitudinal deviations in Tycho Brahe's catalog for stars in the domain A. The a priori date is 1600 AD ($t = 3$).

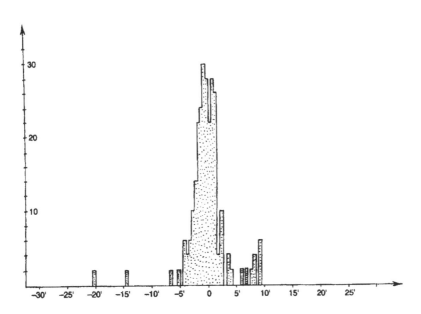

Figure 9.2. Histograms of frequencies of latitudinal deviations in Tycho Brahe's catalog for stars in the domain $Zod\ A$. The a priori date is 1600 AD ($t = 3$).

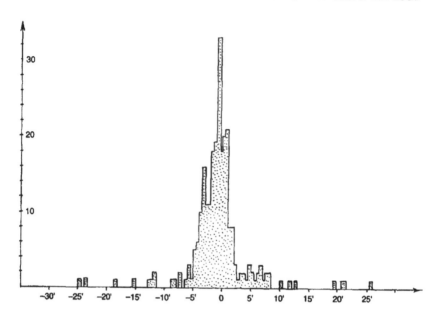

Figure 9.3. Histograms of frequencies of latitudinal deviations in Tycho Brahe's catalog for stars in the domain B. The a priori date is 1600 AD ($t = 3$).

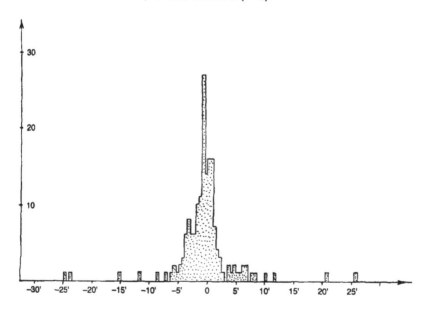

Figure 9.4. Histograms of frequencies of latitudinal deviations in Tycho Brahe's catalog for stars in the domain $Zod\ B$. The a priori date is 1600 AD ($t = 3$).

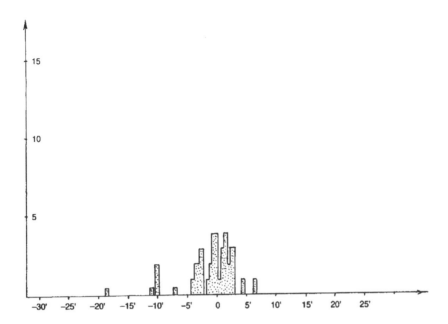

Figure 9.5. Histograms of frequencies of latitudinal deviations in Tycho Brahe's catalog for stars in the domain C. The a priori date is 1600 AD ($t = 3$).

of division here is $0'.5$. Along the vertical, the frequency of occurrence of the error is plotted.

It is obvious from the histograms that outlies occur among latitudinal errors in the catalog of Tycho Brahe. If we assume that the errors in the coordinates are normally distributed (which is natural), then approximately 15% of all errors are outside the "3δ-interval", and so are outlies. Furthermore, the histograms are obviously displaced from zero. The displacement is about $2'$, which indicates that the catalog carries a systematic latitudinal error. The parameter γ that determines the error is approximately equal to $2'$ (see Chapter 5 for the definition of the parameters γ and φ).

In order to weed out the outlies, we excluded from our treatment the stars in the catalog whose latitudinal errors are off the normal distribution (separately in each of the domains A, B, C, D and M). More exactly, we excluded the stars in the domains A, B and M whose latitudinal deviations are more than $5'$ or less than $-7'$, the stars in the domain C whose absolute values of latitudinal deviations exceed $5'$, and the stars in the domain D whose latitudinal deviations are more than $4'$ or less than $-3'$ The bounds had been determined approximately from Figures 9.1–9.7. The total number of excluded stars is 187. The number of the remaining stars, 818, is close to 777,

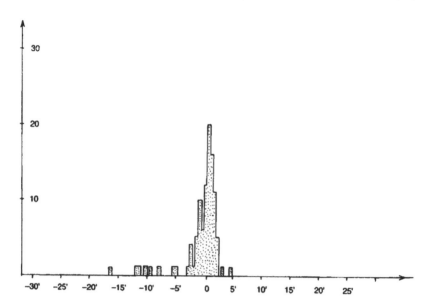

Figure 9.6. Histograms of frequencies of latitudinal deviations in Tycho Brahe's catalog for stars in the domain D. The a priori date is 1600 AD ($t = 3$).

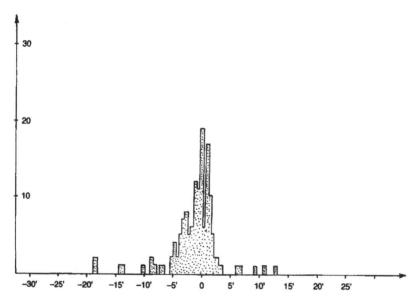

Figure 9.7. Frequencies of latitudinal deviations in Tycho Brahe's catalog for stars in the domain M. The a priori date is 1600 AD ($t = 3$).

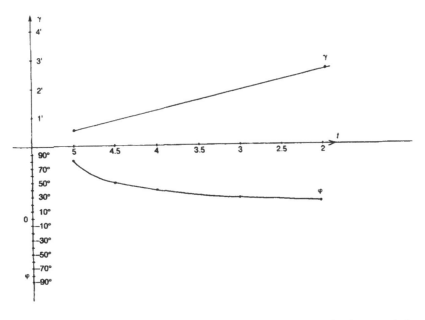

Figure 9.8. Optimal estimates of parameters of systematic errors dependencies on a priori date for Brahe's catalog in the domain A.

the number of stars which, according to a legend (see Ref. 2, p. 126) Tycho Brahe had observed himself.

After the weeding out, we calculated the systematic errors $\gamma_{stat}(t)$ and $\varphi_{stat}(t)$ (see Chapter 5) for the remaining part of the catalog, for each of the seven domains, for the values of t from 5 (1400 AD) to 2 (1700 AD). The results are displayed in Figures 9.8–9.14. It is obvious from the graphs that the values of φ vary from domain to domain, and apparently do not correspond to a systematic error. As for the parameter γ, it has approximately the same value in various domains (note that the situation with the *Almagest* was similar; see Chapter 6). The graphs of $\gamma_{stat}(t)$ dependence for the domains A, Zod A, B, Zod B, C and M are close to each other (Figures 9.8–9.14). The domain D is the only exception: the behavior of γ_{stat} is different in this domain (Figure 9.13). For this reason we did not use the stars in domain D for further dating.

3. Choice of the informative kernel.

Following the suggested algorithm for dating astronomic observations, we need to choose the informative kernel. As indicated in Ref. 3, Tycho Brahe selected 21 stars near the zodiac and determined their equatorial coordinates

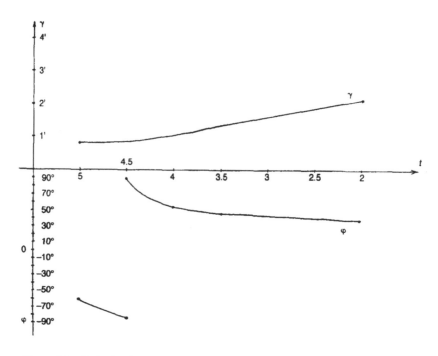

Figure 9.9. Optimal estimates of parameters of systematic errors dependencies on a priori date for the catalog in the domain *Zod A.*

especially carefully (and then recalculated into ecliptic coordinates); the list of these stars is given in Table 9.1.[34]

For the constellations that contain these stars, we have found the group errors $\gamma_{\text{stat}}^G(t)$ and $\varphi_{\text{stat}}^G(t)$ for $t = 3$ (see Section 6.3). We excluded the stars of the constellations G whose group errors $\gamma_{\text{stat}}^G(3)$ differ from $\gamma_{\text{stat}}^{Zod\ A}(3)$ by more than $2'$. For the rest of the constellations, we calculated the rates of stars whose latitudinal deviations do not exceed $1'$, $2'$ and $3'$, and the mean square latitudinal errors, before and after compensation for the systematic error determined by $\gamma = \gamma_{\text{stat}}^G(3)$ and $\varphi = \varphi_{\text{stat}}^G(3)$. Similar characteristics were also calculated after compensation for the systematic error determined by $\gamma = 1.8'$ and $\varphi = 0$. It turned out (see Tables 9.2–9.9) that compensation for the systematic error for each of the constellations considered leads practically to the same result as compensation for the systematic error computed for the collection of the constellations as a whole. Therefore, we may assume that the systematic error with $\gamma = \gamma_{\text{stat}}^{Zod\ A}(t)$, $\varphi = 0$ is the same for all constellations.

We included in the informative kernel twelve stars (of 21) left after weeding out, and also two fast named stars, Arcturus (α Boo) and Procyon (α CMi). The third fast named star, Sirius, was not included in the informative kernel,

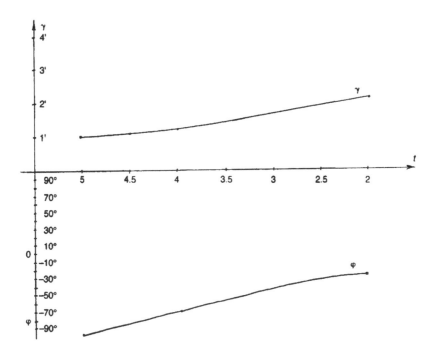

Figure 9.10. Optimal estimates of parameters of systematic errors dependencies on a priori date for the catalog in the domain B.

because it is in the domain D that possesses a specific systematic error (see above). Thus, the informative kernel for the catalog of Tycho Brahe contains 14 stars: γ Ari, α Ari (Hamal), ε Tau, α Tau (Aldebaran), γ Can (Aselli), γ Leo, α Leo (Regulus), γ Vir, α Vir (Spica), δ Oph, α Aqu, α Pis, α Boo (Arcturus), α CMi (Procyon).

4. Dating the observations of Tycho Brahe.

As follows from Tables 9.2–9.9, the mean square latitudinal error (after compensation for the systematic component with $\gamma = \gamma_{\text{stat}}^{\text{Zod}\ A}(t)$, $\varphi = 0$) for the constellations that contain the stars in the informative kernel oscillates between 1′ and 3′, and the rate of stars with individual latitudinal deviations below 2′ is in all cases more than 50%. Following the dating algorithm suggested in Chapter 7, we must put $\Delta = 2$ where Δ is the "claimed accuracy", and find the temporal interval in which the latitudinal errors for all stars in the informative kernel do not exceed $\Delta = 2'$. The ensuing interval is the interval of admissible dates for the catalog of Tycho Brahe.

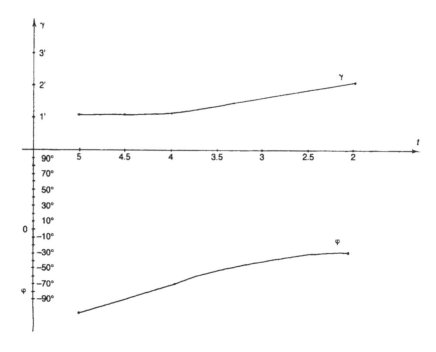

Figure 9.11.   Optimal estimates of parameters of systematic errors dependencies on a priori date for the catalog in the domain *Zod B*.

The resulting interval is 1510 AD–1620 AD (2.8 $\leq t \leq$ 3.9). We used the 10 year step in time; recall that t is counted in centuries back from 1900 AD.

The behavior of maximum (over the informative kernel) latitudinal deviation with variation of the a priori date t is illustrated in Figures 9.15–9.26, the legend of which is similar to that of Figures 7.6–7.23. The domain of parameters (γ, φ) for which the maximum latitudinal error does not exceed 2′ is sparsely hatched, and the domain in which the maximum latitudinal error does not exceed 2′.5 is densely hatched. It is obvious from Figures 9.15–9.26 that the increase of the level of deviation to 2′.5 expands the dating interval only to 1490–1640 (recall that at the level 2′, the interval is 1510–1620). When we set $\Delta = 3'$, we got the interval 1480–1680. Thus, similarly to that for the *Almagest*, the bounds of the dating interval vary but slightly with variation of Δ.

Additional calculations showed that the dating interval for the catalog of Tycho Brahe is stable as well with respect to variations in the choice of the informative kernel.

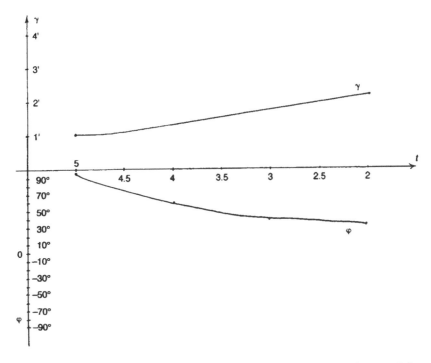

Figure 9.12. Optimal estimates of parameters of systematic errors dependencies on a priori date for the catalog in the domain C.

5. Conclusions.

1) The method for dating astronomic observations applied to the catalog of Tycho Brahe produced the geometrical interval of admissible dates for the catalog 110 years long (1510–1620 AD) that covers the lifetime of Tycho Brahe (1546–1601). The period of observations in the observatory of Uraniborg (1546–1601) is near the middle of this interval.

2) The interval of admissible dates for the catalog is stable with respect to variations of the level Δ and in the choice of the informative kernel. After the increase of Δ from 2′ to 3′ the interval extends to the interval 1480–1680, 200 years long.

3) The ensuing dating interval (110 years) is approximately six times shorter than the interval obtained for the *Almagest*; this agrees with the difference in accuracies of the two catalogs (2′ to 3′ for the catalog of Tycho Brahe in comparison with 10′ to 15′ for the catalog of *Almagest*).

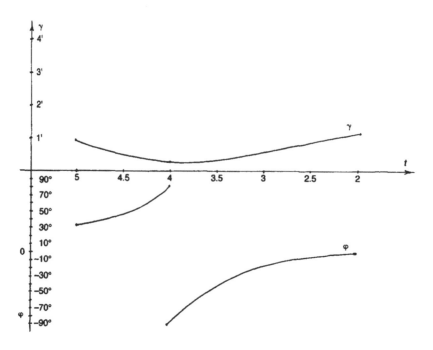

Figure 9.13.   Optimal estimates of parameters of systematic errors dependencies on a priori date for the catalog in the domain D.

4) The statistical interval of admissible dates coincides with the geometrical one for confidence levels $1 - \varepsilon$ that exceed 0.9.

3. Catalog of Ulugh Beg

1. General characteristics of the catalog and results of dating. The catalog of Ulugh Beg is recognized as a refined version of the catalog of the *Almagest*[22], based on astronomic observations undertaken in observatories of Samarkand in the middle of the 15th century (in the reign of king Ulugh Beg). However, as Peters and Knobel note, " ... though in the fifteenth century Ulugh Beg prepared a much more accurate catalogue of Ptolemy's stars, it never came into general use" (Ref. 22, p. 7). This is indeed a catalog of Ptolemy's stars: not only the collection of stars, but also the order in which they are listed are the same in the two catalogs (with very rare exceptions). The number of stars in the catalog of Ulugh Beg is 1019. The ecliptic longitudes and latitudes are given to an accuracy within minutes of arc, but the real accuracy is of course lower (research[2] estimated it as being about 3′ to 5′).

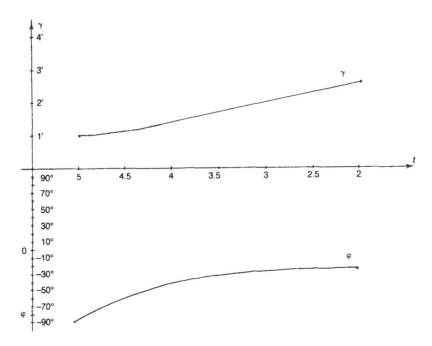

Figure 9.14. Optimal estimates of parameters of systematic errors dependencies on a priori date for the catalog in the domain M.

Our calculations showed that the residual variance of latitudinal errors in the catalog is $16'.5$ for *Zod A*, the best measured domain. Thus, the real accuracy of the catalog is about $30'$ to $35'$—worse than in the *Almagest*!

On the other hand, the value of the systematic error γ for the catalog of Ulugh Beg is less than that for the *Almagest* (see Table 9.10). As a result, the accuracy of latitudes in the original (not compensated) catalog of Ulugh Beg is a little higher (by $5'$ to $6'$) than that of the *Almagest*. However, this difference is immaterial in comparison with the (latitudinal) errors the two catalogs contain in their original form. No wonder that the catalog of the *Almagest* never was ousted by the catalog of Ulugh Beg.

The histogram of latitudinal errors for the catalog is displayed in Figure 9.27; the stars with latitudinal error for $t = 5$ (1400 AD) exceeding $1°$ had been weeded out before plotting this histogram.

The calculations showed also that the catalog contains borrowings from the *Almagest*. Figure 9.28 shows the histogram of differences between the latitudes of stars given in the catalog of Ulugh Beg and the ones in the *Almagest* (identification of stars in the two catalogs presents no difficulties, because, as we have noted, the two catalogs list the stars in the same order). A sharp spike

Table 9.1.

Star	α_{1900} (Ref. 21)	δ_{1900} (Ref. 21)	v_α v_δ (0.001/year; (Ref. 21))	l (Ref. 21)	b (Ref. 28)	magnitude (Ref. 28)
5 γ Ari	$1^h48^m02^s.4$	$+18°48'21''$	$+079$ -108	Υ 27°37'	+7°08'.5	4
13 α Ari	$2^h01^m32^s.0$	$+22°59'23''$	$+190$ -144	$\mho$ 2°06'	+9°57'	3
74 ε Tau	$4^h22^m46^s.5$	$+18°57'31''$	$+108$ -036	$\mathrm{I\!I}$ 2°53'	$-2°36'.5$	3
87 α Tau	$4^h30^m10^s.9$	$+16°18'30''$	$+065$ -189	$\mathrm{I\!I}$ 4°12'.5	$-5°31'$	1
13 μ Gem	$6^h16^m54^s.6$	$+22°33'54''$	$+055$ -112	$\mathrm{I\!I}$ 29°44'	$-0°53'$	3
24 γ Gem	$6^h31^m56^s.1$	$+16°29'05''$	$+043$ -044	$\mathfrak{S}$ 3°31'	$-6°48'.5$	2
78 β Gem	$7^h39^m11^s.8$	$+28°16'04''$	-627 -051	$\mathfrak{S}$ 17°43'	+6°38'	2
43 γ Can	$8^h37^m29^s.9$	$+21°49'42''$	-103 -043	Ω 1°57'	+3°08'	4
41 γ Leo	$10^h14^m27^s.6$	$+20°20'51''$	$+307$ -151	Ω 23°59'	+8°47'	2
32 α Leo	$10^h03^m02^s.8$	$+12°27'22''$	-249 -003	Ω 24°17'	+0°26'.5	1
29 γ Vir	$12^h36^m35^s.5$	$-0°54'03''$	-568 -008	$\simeq$ 4°35'.5	+2°50'	3
67 α Vir	$13^h19^m55^s.4$	$-10°38'22''$	-043 -033	$\simeq$ 18°16'	$-1°59'$	1
27 β Lib	$15^h11^m37^s.4$	$-9°00'50''$	-098 -023	M 13°48'	+8°35'	2
1 δ Oph	$16^h19^m06^s.2$	$-3°26'13''$	-048 -145	M 26°44'.5	+17°19'	3
21 α Sco	$16^h23^m16^s.4$	$-26°13'26''$	-007 -023	$\nearrow$ 4°13'	$-4°27'$	1
39 o Sag	$18^h58^m41^s.4$	$-21°53'17''$	$+079$ -060	$\mathrm{る}$ 9°28'	+0°59'	4
53 α Aqi	$19^h45^m54^s.2$	$+8°36'15''$	$+537$ $+385$	$\mathrm{る}$ 26°09'	+29°21'.5	2
40 γ Capr	$21^h34^m33^s.1$	$-17°06'51''$	$+188$ -022	$\approx$ 16°14'	$-2°26'$	3
22 β Aqu	$21^h26^m17^s.7$	$-6°00'40''$	$+019$ -005	$\approx$ 17°51'	+8°42'	3
54 α Peg	$22^h59^m46^s.7$	$+14°40'02''$	$+062$ -038	$\mathrm{\times}$ 17°56'.5	+19°26'	2
113 α Pis	$1^h56^m52^s.3$	$+2°16'51''$	$+030$ 000	Υ 23°47'.5	$-9°04'.5$	3

at the zero in Figure 9.28 corresponds to the group of stars whose latitudes in the two catalogs are exactly the same; the height of the spike leaves no place for coincidence by chance.

Below, we obtain the geometric interval of admissible dates for the observations of Ulugh Beg. This is the interval 700 AD–1450 AD. The interval covers the traditionally accepted date of observations of Ulugh Beg (the first half of the 15th century). The length of the interval amounts to 650 years, and is close to the length of the dating interval for the *Almagest* (600 AD–1300 AD). This is quite natural, for the two catalogs have similar accuracy. Note also that the dating intervals for the two catalogs overlap strongly, so it is possible that they were compiled in approximately the same time.

Table 9.2. Cancer (13 stars)

	The rate of stars whose latitudinal error does not exceed			Residual mean square error
	$1'$	$2'$	$3'$	$\hat{\sigma}$
Initial	38%	77%	77%	$2\rlap{.}'4$
After the turn appropriate to Zod A	61%	85%	92%	$2\rlap{.}'37$
After the optimal turn for this constellation	61%	77%	92%	$2\rlap{.}'31$
After the turn with $\gamma = \gamma_{\text{stat}}^{Zod\ A}, \varphi = 0$	46%	77%	92%	$2\rlap{.}'77$

Table 9.3. Leo (36 stars)

	The rate of stars whose latitudinal error does not exceed			Residual mean square error
	$1'$	$2'$	$3'$	$\hat{\sigma}$
Initial	61%	83%	94%	$1\rlap{.}'41$
After the turn appropriate to Zod A	55%	80%	94%	$1\rlap{.}'44$
After the optimal turn for this constellation	61%	83%	94%	$1\rlap{.}'35$
After the turn with $\gamma = \gamma_{\text{stat}}^{Zod\ A}, \varphi = 0$	47%	75%	94%	$1\rlap{.}'63$

Table 9.4. Taurus (37 stars)

	The rate of stars whose latitudinal error does not exceed			Residual mean square error
	$1'$	$2'$	$3'$	$\hat{\sigma}$
Initial	76%	89%	94%	$1\rlap{.}'18$
After the turn appropriate to Zod A	54%	92%	97%	$1\rlap{.}'31$
After the optimal turn for this constellation	67%	92%	94%	$1\rlap{.}'17$
After the turn with $\gamma = \gamma_{\text{stat}}^{Zod\ A}, \varphi = 0$	24%	62%	94%	$1\rlap{.}'94$

Table 9.5. Pisces (31 stars)

	The rate of stars whose latitudinal error does not exceed			Residual mean square error
	1'	2'	3'	$\hat{\sigma}$
Initial	61%	77%	90%	1.'81
After the turn appropriate to Zod A	48%	81%	90%	1.'97
After the optimal turn for this constellation	64%	81%	90%	1.'79
After the turn with $\gamma = \gamma_{stat}^{Zod\ A}$, $\varphi = 0$	45%	77%	87%	1.'87

Table 9.6. Aquarius (34 stars)

	The rate of stars whose latitudinal error does not exceed			Residual mean square error
	1'	2'	3'	$\hat{\sigma}$
Initial	29%	56%	76%	2.'49
After the turn appropriate to Zod A	32%	59%	82%	2.'23
After the optimal turn for this constellation	35%	82%	91%	1.'63
After the turn with $\gamma = \gamma_{stat}^{Zod\ A}$, $\varphi = 0$	38%	65%	91%	1.'9

Table 9.7. Virgo (32 stars)

	The rate of stars whose latitudinal error does not exceed			Residual mean square error
	1'	2'	3'	$\hat{\sigma}$
Initial	25%	72%	94%	1.'8
After the turn appropriate to Zod A	34%	72%	94%	1.'83
After the optimal turn for this constellation	62%	91%	100%	1.'16
After the turn with $\gamma = \gamma_{stat}^{Zod\ A}$, $\varphi = 0$	59%	91%	94%	1.'26

Table 9.8. Aries (20 stars)

	The rate of stars whose latitudinal error does not exceed			Residual mean square error
	$1'$	$2'$	$3'$	$\hat{\sigma}$
Initial	65%	85%	100%	$1\!.22$
After the turn appropriate to Zod A	60%	40%	100%	$1\!.21$
After the optimal turn for this constellation	50%	95%	100%	$1\!.20$
After the turn with $\gamma = \gamma_{\text{stat}}^{\text{Zod }A}, \varphi = 0$	45%	65%	90%	$1\!.63$

Table 9.9. Ophiuchus (24 stars)

	The rate of stars whose latitudinal error does not exceed			Residual mean square error
	$1'$	$2'$	$3'$	$\hat{\sigma}$
Initial	17%	37%	70%	$2\!.84$
After the turn appropriate to Zod A	46%	79%	92%	$1\!.93$
After the optimal turn for this constellation	50%	92%	92%	$1\!.69$
After the turn with $\gamma = \gamma_{\text{stat}}^{\text{Zod }A}, \varphi = 0$	25%	54%	83%	$2\!.40$

2. Systematic errors in the catalog of Ulugh Beg.

The values of the parameters for the systematic errors $\gamma_{\text{stat}}(t)$ and $\varphi_{\text{stat}}(t)$ (see Section 2.3 for definition) were found for the domain Zod A for $1 \le t \le 20$ (100 BC–1800 AD); the values for $t = 4$ (1500 AD), $t = 10$ (900 AD) and $t = 15$ (400 AD) are shown in Table 9.10. The table contains also the values of the mean square error $\hat{\sigma}$ for the original coordinates and after compensation for the systematic error with $\gamma = \gamma_{\text{stat}}$ and $\varphi = \varphi_{\text{stat}}$.

3. Choice of the informative kernel and of the level Δ. Dating the catalog.

Similarly to what we did as we dated the catalog of the *Almagest*, we included in the informative kernel for the catalog of Ulugh Beg the nine named

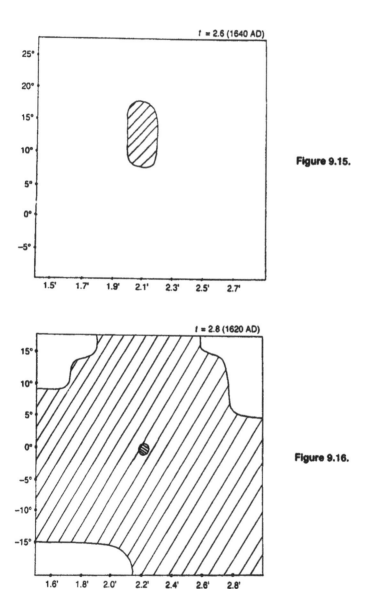

Figure 9.15.

Figure 9.16.

Figures 9.15 through 9.26. Dating Tycho Brahe's catalog. The figure shows the minimax deviation $\Delta(t, \gamma, \varphi)$ dependence on all the three variables, the a priori date t and the parameters of systematic error γ and φ. At the fixed value of t the rectangular confidence neighborhood of $(\gamma_{\text{stat}}^{Zod\ A}(t), \varphi_{\text{stat}}^{Zod\ A}(t))$ is shown in the plane (γ, φ). The densely hatched domain contains the values of parameters that provide minimax deviation below $2'$. The domain where the deviation does not exceed $2.5'$ is hatched.

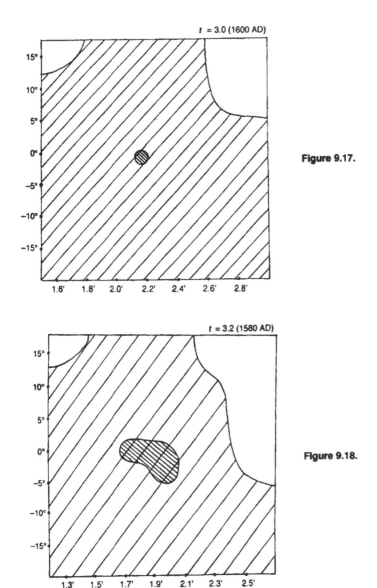

Figure 9.17.

Figure 9.18.

Figures 9.15 through 9.26. Dating Tycho Brahe's catalog. The figure shows the minimax deviation $\Delta(t, \gamma, \varphi)$ dependence on all the three variables, the a priori date t and the parameters of systematic error γ and φ. At the fixed value of t the rectangular confidence neighborhood of $(\gamma_{stat}^{Zod}{}^A(t), \varphi_{stat}^{Zod}{}^A(t))$ is shown in the plane (γ, φ). The densely hatched domain contains the values of parameters that provide minimax deviation below $2'$. The domain where the deviation does not exceed $2.5'$ is hatched.

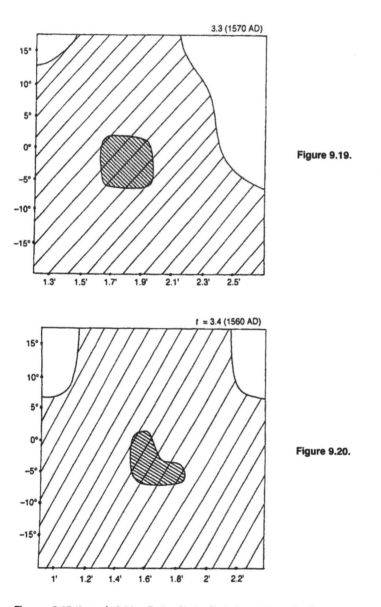

Figure 9.19.

Figure 9.20.

Figures 9.15 through 9.26. Dating Tycho Brahe's catalog. The figure shows the minimax deviation $\Delta(t, \gamma, \varphi)$ dependence on all the three variables, the a priori date t and the parameters of systematic error γ and φ. At the fixed value of t the rectangular confidence neighborhood of $(\gamma_{\text{stat}}^{Zod\ A}(t), \varphi_{\text{stat}}^{Zod\ A}(t))$ is shown in the plane (γ, φ). The densely hatched domain contains the values of parameters that provide minimax deviation below $2'$. The domain where the deviation does not exceed $2.5'$ is hatched.

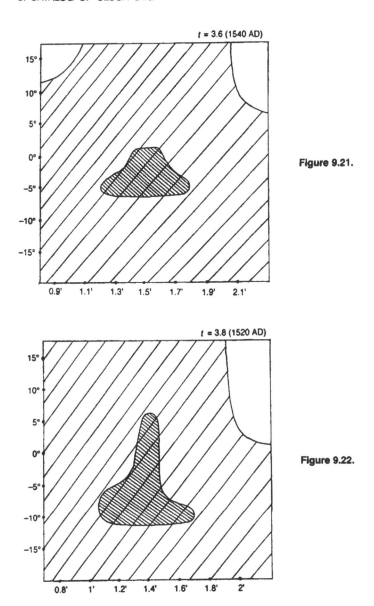

Figure 9.21.

Figure 9.22.

Figures 9.15 through 9.26. Dating Tycho Brahe's catalog. The figure shows the minimax deviation $\Delta(t, \gamma, \varphi)$ dependence on all the three variables, the a priori date t and the parameters of systematic error γ and φ. At the fixed value of t the rectangular confidence neighborhood of $(\gamma_{\text{stat}}^{Zod\ A}(t), \varphi_{\text{stat}}^{Zod\ A}(t))$ is shown in the plane (γ, φ). The densely hatched domain contains the values of parameters that provide minimax deviation below $2'$. The domain where the deviation does not exceed $2.5'$ is hatched.

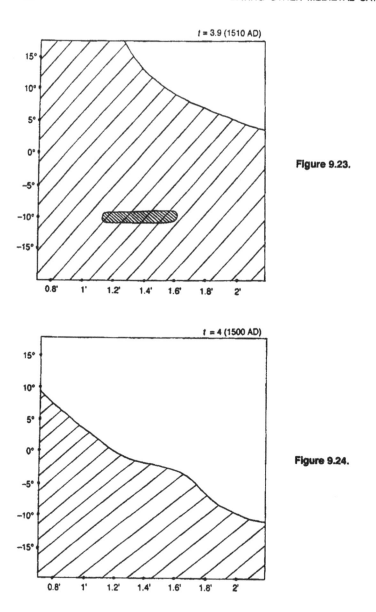

Figure 9.23.

Figure 9.24.

Figures 9.15 through 9.26. Dating Tycho Brahe's catalog. The figure shows the minimax deviation $\Delta(t, \gamma, \varphi)$ dependence on all the three variables, the a priori date t and the parameters of systematic error γ and φ. At the fixed value of t the rectangular confidence neighborhood of $(\gamma_{stat}^{Zod}{}^A(t), \varphi_{stat}^{Zod}{}^A(t))$ is shown in the plane (γ, φ). The densely hatched domain contains the values of parameters that provide minimax deviation below $2'$. The domain where the deviation does not exceed $2.5'$ is hatched.

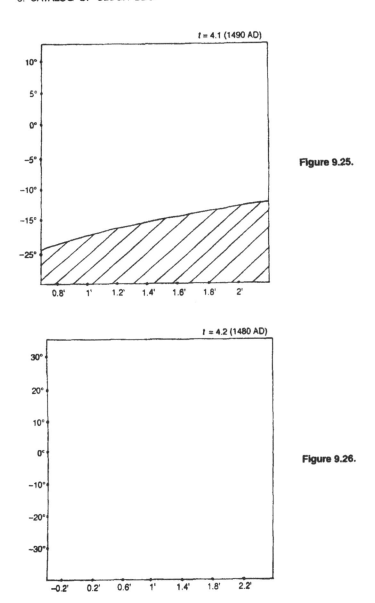

$t = 4.1$ (1490 AD)

Figure 9.25.

$t = 4.2$ (1480 AD)

Figure 9.26.

Figures 9.15 through 9.26. Dating Tycho Brahe's catalog. The figure shows the minimax deviation $\Delta(t, \gamma, \varphi)$ dependence on all the three variables, the a priori date t and the parameters of systematic error γ and φ. At the fixed value of t the rectangular confidence neighborhood of $(\gamma_{\text{stat}}^{Zod}{}^{A}(t), \varphi_{\text{stat}}^{Zod}{}^{A}(t))$ is shown in the plane (γ, φ). The densely hatched domain contains the values of parameters that provide minimax deviation below $2'$. The domain where the deviation does not exceed $2.5'$ is hatched.

Table 9.10.

	γ_{stat}	φ_{stat}	σ_{init}	σ_{min}
$t = 4$	11.55	−43°	18.36	16.43
$t = 10$	10.33	−60°	17.92	16.33
$t = 15$	10.87	−76°	18.10	16.35

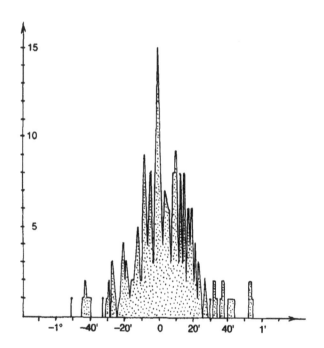

Figure 9.27.   Ulugh Beg catalog. Histogram of frequencies of latitudinal errors in the domain A.

stars that lie in the domain A, the best measured part of the sky, or near. The nine stars are exactly the ones that are named in the *Almagest*: Arcturus (α Boo), Regulus (α Leo), Spica (α Vir), Antares (α Sco), Capella (α Aur), Lyra (Vega; α Lyr), Aselli (γ Can), Procyon (α CMi) and Vindemiatrix (ε Vir). This time we do not exclude Vindemiatrix (as we did for the *Almagest*), because its coordinates in the catalog are not a result of a later calculation and are, apparently, not a copyist's error[28].

According to Table 9.10, we must choose as the true accuracy of the catalog Δ the value 10′, which coincides with the true accuracy of the catalog of the *Almagest*. Indeed, after compensation for the systematic error, the mean

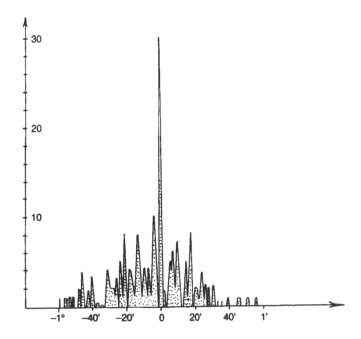

Figure 9.28. Ulugh Beg catalog compared with the *Almagest*. Histogram of frequencies of differences between latitudes of stars as given in Ulugh Beg catalog and in the *Almagest*. The sharp peak at zero tells that Ulugh Beg borrowed some coordinates from the *Almagest*.

square latitudinal error for *Zod A* is equal to 16′.5, and 45% of stars in the domain have latitudinal errors below 10′ (after compensation for the systematic error).

The choice of the above informative kernel and of the value $\Delta = 10'$ leads to the interval of admissible dates 700 AD–1450 AD. The statistical interval of admissible dates coincides with the geometrical one at the values of the confidence level above 0.4.

The dating interval is stable with respect to variations of Δ and of the choice of the informative kernel. Thus, at $\Delta = 15'$ the interval expands to 400 AD–1600 AD.

The dependence of the minimax latitudinal deviation over the informative kernel $\delta(t)$ is shown in Figure 9.29. The graph is similar to the graph in Figure 7.27 that displays the similar dependence for the catalog of the *Almagest*. Recall that $\delta(t)$ is the minimum over all possible matchings of the configuration of the stars in the informative kernel with the corresponding true (computed) configuration in the year t of the maximum latitudinal error over all stars in the informative kernel (it is clear how to calculate individual

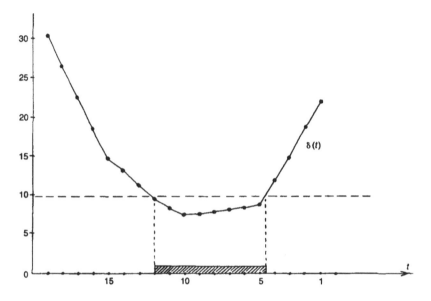

Figure 9.29. Ulugh Beg catalog. Minimax latitudinal deviation dependence on the a priori date. The ensuing interval of admissible dates for the catalog is 700–1450 AD.

latitudinal errors for each star, and hence their maximum, having a matching fixed). Figure 9.29, in particular, exhibits the dependence of the dating interval for the catalog of Ulugh Beg on the level Δ. A comparison of Figures 9.29 and 7.27 confirms similarity of the accuracy characteristics of the two catalogs.

4. Conclusions.

1) The geometrical dating interval for the catalog of Ulugh Beg is 700–1450 AD. The interval covers the traditionally accepted date of compilation of the catalog (1437 AD), although this date is near the upper bound of the interval. On the other hand, the interval is very close to the similar interval for the *Almagest*, so it is possible that the dates of compilation of the two catalogs are close to each other.

2) The accuracy characteristics of the catalog of Ulugh Beg and the catalog of the *Almagest* are quite similar; however, there are some minor differences: the systematic latitudinal error in the *Almagest* is greater than Ulugh Beg's (about 20′ in the *Almagest*, and about 10′ in Ulugh Beg's catalog), but the random errors are a little greater in Ulugh Beg's catalog ($\hat{\sigma} = 16′.5$; in the *Almagest*, $\sigma = 12′.8$). The same coordinates are given in the two catalogs for 48 stars, apparently, the result of a direct borrowing.

3) The dating interval for the catalog of Ulugh Beg is stable with respect to variations of the level Δ and of the choice of the informative kernel.

4) At the values of the confidence level above 0.4, the statistical interval of admissible dates coincides with the geometrical. If we raise the level Δ to 15′, the corresponding statistical interval for $\varepsilon \leq 0.01$ narrows by 100 years down from the "geometric" upper bound, and only reaches 1500 AD.

4. Catalog of Hevelius

The catalog of Hevelius was compiled in the 17th century, after invention of the telescope. However, Hevelius did not use telescopes, because he thought that observations with an unaided eye were more exact than telescopic[28]. This was confirmed by Halley after a "competition" with Hevelius, in which they measured coordinates of the same stars in different ways: Halley used a telescope and Hevelius used traditional astronomic instruments. The results allegedly differed by no more than 1″. Since then, it is accepted that the accuracy of coordinates in the catalog of Hevelius is close to the accuracy of telescopic observations of the time, about 1″ (the coordinates are given in the catalog to seconds of arc).

Our analysis does not confirm this. We considered several configurations of bright named stars in the catalog, including three fast stars: Arcturus (α Boo), Sirius (α CMa) and Procyon (α CMi). Varying t in the a priori dating interval $1 \leq t \leq 5$, we tried to find for each value of t a matching of the stellar configuration in the catalog with the corresponding true (computed) configuration so as to minimize the maximum latitudinal deviation for the configuration (as usual, we mean *ecliptic* latitudes). The parameters that provide this optimal matching turned out to be $\gamma = 0$ and $\varphi = 0$, which means that the configurations in question carry no systematic errors in the catalog of Hevelius. As for random errors, they are about as large as in the catalog of Tycho Brahe (2′ to 3′). The division value in the catalog of Hevelius is 60 times less than in the catalog of Tycho Brahe, so latitudinal errors in the catalog of Hevelius are 100 to 200 times the division value!

Figure 9.30 illustrates this. Here, we display latitudinal errors as functions of the a priori date t for each of the ten bright named stars in the catalog of Hevelius, Arcturus (α Boo), Sirius (α CMa), Procyon (α CMi), Antares (α Sco), Lyra (Vega; α Lyr), Pollux (β Gem), Castor (α Gem), Spica (α Vir), Capella (α Aur) and Regulus (α Leo). Figure 9.31 shows similar dependencies found for the catalog of Tycho Brahe. A comparison of the two figures shows similarity of general character of errors in the two catalogs, as well as of individual latitudinal errors of several stars (Arcturus, Sirius, Antares, Procyon and Lyra). This indicates a dependence between the two catalogs.

Let us formulate our conclusions.

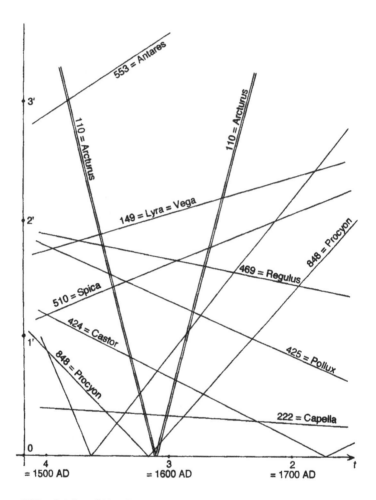

Figure 9.30. Catalog of Hevelius. Individual latitudinal errors for ten bright stars in the catalog as functions of the a priori date.

1) The accuracy of the Hevelius' catalog is not higher than the accuracy of the Brahe's catalog. This follows from an analysis of configuration of bright named stars in the catalog of Hevelius.

2) The catalog of Hevelius is apparently not independent of the catalog of Tycho Brahe. The dependence is most pronounced in the group of fast bright stars (Arcturus, Sirius and Procyon). Since an analysis of fast stars is what our dating method is based on, an application of the method to the catalog of Hevelius is senseless; the ensuing date will be similar to the result of dating Tycho Brahe's catalog.

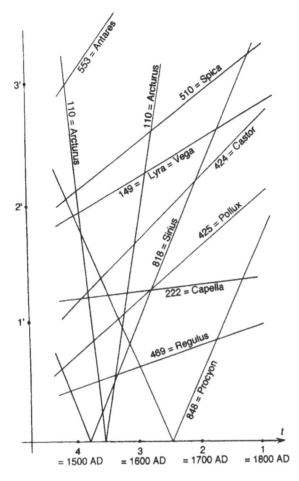

Figure 9.31. Tycho Brahe's catalog compared with the catalog of Hevelius. Individual latitudinal deviations for the same ten stars as in Figure 9.30. The comparison shows that the accuracy of the catalog of Hevelius is not higher than that of Tycho Brahe's catalog (in contradiction with a common opinion).

5. About Al Sûfi's catalog

We considered the copy of Al Sûfi's catalog adduced in Ref. 57. It is usually considered that the catalog was compiled by Al Sûfi from his own observations [56,57]. Al Sûfi, in his introduction to the catalog, opposes those astronomers who compile star catalogs under their names not having observed real stars, but using catalogs compiled by their predecessors (the *Almagest*) and stellar globes: "I have seen many those questing for knowledge of fixed stars ... and I found two categories of these men.

"The first, following methods of astronomers, use globes painted by artists who, not having their own knowledge about stars, only take longitudes and latitudes in books and place the stars in the sphere, unable to distinguish truth from mistakes. And skilled men, as they look at the globes, see that many stars are placed there not in the way they are in the sky. Those who make the globes refer to astronomic tables whose authors pretend that they had observed the stars and found their positions. And in fact, they only chose the most famous stars, which all know, like the Eye of Aries (Aldebaran— *Authors*), Lion's Heart (Regulus — *Authors*), Virgo's Spike (Spica—*Authors*), three stars in the brow of Scorpio and his Heart (Antares—*Authors*), of which Ptolemy tells that he had observed their longitudes and latitudes, given in the *Almagest*, because these stars are close to the ecliptic. As for the other stars marked by Ptolemy in the star catalog of his book, they add to each of them what they think is needed. Having moved these stars in space for the time between their lives and the life of Ptolemy, they add several minutes more to the longitudes of Ptolemy, or subtract, in order to make believe that they had observed themselves and found some differences in latitudes and longitudes independently of their common increment in the time that elapsed between them and Ptolemy. All this they did, not even knowing the stars themselves. Among those are Al Bāttāni and Atarid, and other.

"I studied carefully many copies of the *Almagest* and found that they differ in what concerns many fixed stars ... The second category of those questing for the knowledge of fixed stars are amateurs" (quoted from Ref. 4, vol. 4). However, it obviously follows from a comparison of stellar coordinates given in the *Almagest* and in the Al Sûfi's catalog that the catalog of Al Sûfi is nothing more but one of various versions of the *Almagest*. Indeed, the order of stars in the two catalogs is exactly the same; the longitudes of all stars are by $12°42'$ more than that in the canonical version of the *Almagest*, and latitudes are exactly the same. Note that a similar shift of longitudes by a constant (that is, reduction for precession for a different epoch) can be found in many copies of the *Almagest* (see Ref. 22).

A comparison of latitudes of all stars in the catalog of Al Sûfi[57] and in the *Almagest* [22] showed that only 53 stars of 1028 have different latitudes in the two catalogs. This number of differences is usual for various versions of the *Almagest*. Moreover, latitudes of 35 of the 53 stars occur in the versions of the *Almagest* investigated by Peters and Knobel[22]. Thus, the catalog of Al Sûfi is nothing more than another version of the catalog of the *Almagest*.

Figure 9.32 displays the diagram which shows for each of the 25 manuscripts of the *Almagest* investigated in Ref. 22 the number of stars whose latitudes in the manuscript differ from the ones given in the canonical version of the *Almagest* by the same value as do the latitudes given in the catalog of Al Sûfi

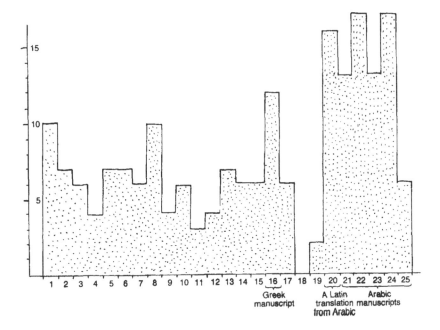

Figure 9.32. Al Sûfi's catalog compared with manuscripts of the *Almagest* treated by Peters and Knobel. The diagram shows the number of cases when the latitude of a star given by Al Sûfi differs from the one given in the canonical version of the *Almagest* by exactly as much as the latitude given in the manuscript of the *Almagest* differs from the one given in the canonical version of the *Almagest*.

(i.e. as much as the latitudes given by Al Sûfi differ from the ones given in the canonical version of the *Almagest*). It is interesting to note that the group of manuscripts of the *Almagest* most close to the catalog of Al Sûfi (manuscripts 20–24 in Figure 9.32) consists of Arabic manuscripts related to a common protograph, the so-called "translation of Al Mamon", the translation of the *Almagest* made, presumably, by Al Mamon in the 9th century—see Ref. 22, p. 23. Apparently, the catalog of Al Sûfi should be attributed to this group of versions of the *Almagest*.

In conclusion, we adduce the words of Peters and Knobel: "The translation into French from the Arabic of Abd Al Rahman Al Sûfi, by Shjellerup, is simply Ptolemy's catalogue for a different epoch" (Ref. 22, p. 7).

Chapter 10
A Date for the *Almagest* from Coverings of Stars and Lunar Eclipses

1. Introduction

The date for the catalog of the *Almagest* we have found in the previous chapters from geometrical and statistical analysis of latitudes contradicts sharply the date the *Almagest* is traditionally attributed to (137 AD); however, testing the method on several catalogs whose dates of compilation raise no doubts, as well as on artificially produced catalogs confirms its efficiency. So, the following question naturally arises: Is it possible that the star catalog is a later insertion in the authentic ancient text of the *Almagest*, or was the text of the *Almagest* (or its major part) written about the same time as the catalog (not before 600 AD)?

The astronomic data presented in the catalog have been recently investigated (very thoroughly and professionally) by the well-known American astronomer, specialist in celestial mechanics, navigation and astrophysics Robert R. Newton[1]. His conclusions may be summarized as follows:

1) The *Almagest* contains elements of theory of lunar motion, solar motion, planetary motion and of precession.

2) Much astronomic data presented in the *Almagest* (in particular, many "observations") could have been calculated on the basis of Ptolemy's theory.

3) Many astronomic "observations" are indeed nothing more than the result of these "purely theoretical calculations", made by Ptolemy.

Consequently, a use of these data for dating the *Almagest* can only lead to a reconstruction of the opinion (or the conjecture) of Ptolemy (or some other medieval astronomer?) about the time when these or that astronomic events really occurred. Medieval astronomers sometimes solved problems about the particular month when some concrete astronomic events took place in a given year in far the past.

Fortunately, the *Almagest* contains some descriptions of astronomic events that could not be calculated from late medieval astronomic theories, not to mention from Ptolemy's theory. First of all, these are ecliptic latitudes of 1028 stars contained in the catalog; we used these data for dating the catalog in the previous chapters of this book. Among other data that could not be obtained by mere calculation are

A) four observations of coverings of stars by planets, and
B) 21 observations of lunar eclipses.

In this chapter, we date these events, thus finding a date for the *Almagest*; we should stress that in this case we deal with the *text* of the *Almagest*, but not with the star catalog.

We obtained the following:

1) The events of the group A (coverings of stars by planets) could take place in the interval 887 AD–1009 AD. This interval agrees perfectly with the dating interval for the star catalog obtained in Chapter 7.

2) According to the *Almagest*, the lunar eclipses that form the group B were distributed in a 900 years long interval; our estimate for this interval is 492 AD–1350 AD. Moreover, the eclipses occurred most frequently in the 11th century. Again, the dating interval agrees perfectly with the one found in Chapter 7 from the star catalog.

3) Ptolemy attributed observations in both groups to the same "era", called *the era of Nabonassar*. Now as we have the dates for the observations, we may date the starting point of this era from two independent sets of data. It is remarkable that the two resulting dates coincide: *the date for the beginning of the Nabonassar era is about 490 AD*. Recall that the traditional date for the beginning of this era is 747 BC.

It is important that the three sets of astronomic data presented in the *Almagest*, the latitudes in the star catalog, the data about coverings of stars by planets, and observations of lunar eclipses are *completely independent*, so the agreement of the dates obtained from the three sets is a strong argument for the *Almagest* being originally written in the 10th–11th centuries, and extended and supplemented about the middle of the 14th century.

2. A date from coverings of stars by planets

The *Almagest* contains descriptions of four coverings of stars by planets[21,37]. Ptolemy tells:

1) "Of the old observations, we took one which Timocharis record thus: In the year 13 of Philadelphus, Egyptianwise Mesore 17–18 at the twelfth hour, Venus appeared to have exactly overtaken the star opposite Vindemiatrix" (Ref. 24, p. 319; Section X.4).

2) "We took one of the old observations to which it is quite clear that in the year 13 according to Dionysius, Aigon 25 in the morning, Mars seemed to occult the Scorpius' northern forehead" (Ref. 24, p. 342, Section X.9).

3) "We again took one of the ancient observations very faithfully recorded, according to which it is quite clear that in the year 45 of Dionysius, Parthenon 10, Jupiter at sunrise occulted the Southern Ass" (Ref. 24, p. 361; Section XI.3).

4) "We took for this again one of the faithfully recorded ancient observations, according to which it is clear that in the year 82 of the Chaldeans, Xanthicus 5, in the evening, Saturn was 2 digits below the Virgin's southern shoulder" (Ref. 24, p. 379, Section XI.7).

If we use the traditional identifications of Ptolemy's stars with the modern ones[1,37], we have here descriptions of the following events:

1) Venus covered the star η Virgo about midnight;
2) Mars covered the star β Scorpius in the morning;
3) Jupiter covered the star δ Cancer at sunrise.
4) Saturn was "2 digits" (two units?) below the star γ Virgo.

We checked the traditional identifications, and found them reliable. For calculation of planetary positions, we used a modern theory and concrete values of the averaged elements of planetary orbits given in Ref. 38; see Appendix 1 at the end of this chapter. The accuracy of calculation of latitudes is $1'$ (one minute of arc).

Let us now discuss the meaning of the words *a planet covered a star*.

It is well known that an ordinary human eye can distinguish two points separated by angular distance of $1'$. Extremely strong eyes can distinguish two points about $30''$ distant from each other. Consequently, the covering of a star by a planet means in reality that the angular distance between them does not exceed $1'$ (or even $30''$). Clearly, it was impossible for Ptolemy to calculate a covering, because the accuracy of any measurement he had at his disposal never was better than $10'$. Modern theory enables us to calculate latitudes of Mars and Venus to the past, within the temporal interval we are interested in, to an accuracy $1'$; as for longitudes, the accuracy of their calculation is $3'$.

This is sufficient for our purposes, for in fact the latitude alone determines a covering of a star by a planet. The longitudes of planets vary rapidly (in comparison with latitudes), and we may assume that longitudes are linear functions of time. Consequently, a small longitudinal error introduces but a small error in the date of a covering. Thus, the coverings by Mars and Venus may be calculated with high accuracy from the modern theory.

The theory of Jupiter and Saturn is more complicated and less accurate. Concerning this, V. K. Abalakin writes: "The averaged elements of the orbits of Jupiter, Saturn, Uranus, Neptune and Pluto cannot be used for solving the stability problem and are not valid in millions years long intervals of time ... They are only valid for several centuries from our time" (Ref. 39, p. 302).

But in the case of the *Almagest*, we do not need precise formulas for Jupiter and Saturn. Indeed, it is obvious from the above description that the observation of Saturn is but auxiliary, because Saturn did not cover the star, but was "two digits below" the star. What Ptolemy meant here under the term "digit" is not clear. Therefore it is senseless to calculate the position of Saturn to an accuracy of $1'$.

In the case of Jupiter, Ptolemy tells that "Jupiter occulted the star". Our computer calculation shows that the distance between Jupiter and δ Cancer has never been less than $15'$ in the temporal interval we are interested in. So, we are only left to find the moments when the distance between Jupiter and the star was $15'$ to $20'$, and to that end no high accuracy is required; the accuracy guaranteed by modern theory is quite sufficient.

Let us now look at how Ptolemy distributes the events 1)–4) in the time axis. Ptolemy uses the "era of Nabonassar"; usually he attributes the events to the years of this era, though other epochs are also sometimes in use; Ptolemy also uses years "after the death of Alexander" and "according to Dionysius". Thus, the event 1) is attributed to the year 406 of the era of Nabonassar, the event 2) to the year 476 of Nabonassar, year 42 after the death of Alexander, and to the year 13 according to Dionysius, the event 3) to the year 83 after the death of Alexander and the year 5 according to Dionysius; finally, the event 4) to the year 519 of Nabonassar. It is obvious that these data are contradictory: the time between the events 2) and 3) is 41 years if we use the era of Alexander, and is 32 years if we use the era of Dionysius. This implies two dates in the years of Nabonassar: year 517 and year 508. Below, we consider both versions.

Thus, we are now ready to state an exact mathematical problem. The problem is to find the year N with the following properties:

(1) In the year N, Venus covered η Virgo about midnight;
(2) In the year $N + 70$, Mars covered β Scorpius in the morning;
(3) In the year $N + 111$ or $N + 102$, Jupiter covered δ Cancer at sunrise;
(4) In the year $N + 113$, Saturn was near ("two digits below") γ Virgo.

Let us now look at what accuracy of the dates we may expect. The fact is that we must take into account the errors introduced by Ptolemy's recalculation of all dates to the same era (of Nabonassar) due to the fact that different eras used various starting months for the years. So, it would not be surprising to get an error amounting to several years. The best solution we have found has a 4 years error.

Assertion 1. *The above problem has only one solution in the interval 500 BC– 1600 AD. The solution is $N = 887$ AD:*

(1) *On September 9, 887 AD, at midnight of Greenwich time, Venus was at the distance within 1' from η Virgo;*

(2) *On January 27, 959 AD, at 6:50 Greenwich, Mars was at the distance about 3' from β Scorpius;*

(3) *On August 13, 994 AD, at 5:13 Greenwich, the distance between Jupiter and δ Cancer was about 20'. This distance is about the minimum possible distance between Jupiter and the star in the interval considered;*

(4) *On September 30, 1009 AD, at 4:50 Greenwich, Saturn was at the distance 50' from γ Virgo (below the star).*

Remark. There exists also an "antique" solution, which strictly speaking is not a solution, because one of the deviations makes eight years; however, the rest of the deviations do not exceed four years, so we have an "almost solution":

(1) On September 1, 329 BC, 19:45 Greenwich, Venus was at the distance within 1' from η Virgo;

(2) On January 17, 257 BC, 5:10 Greenwich, Mars was at the distance within 1' from β Scorpius;

(3) On September 9, 229 BC, 4:15 Greenwich, Jupiter was at the distance about 15' from δ Cancer; the distance is near the minimum possible distance between Jupiter and the star in the interval considered.

(4) On September 6, 229 BC, 15:10 Greenwich, Saturn was at the distance 127' from γ Virgo (below the star).

For both solutions, the deviations of the intervals between the successive events from the intervals indicated by Ptolemy do not exceed 4 years. If we exclude observations of Saturn (event (4)), then the maximum deviation of the first solution is three years. No other solutions with deviations below 10 years exist.

All dates are Julian, with the years starting on January 1. In the sequel we call the solution described in Assertion 1 *medieval*, and the "almost solution" *antique*.

There is a "solution" of the problem accepted as such by chronologists of the 16th–17th centuries (see Ref. 1); it suggests $N = 272$ BC:

(1) October 12, 272 BC. Venus "touched" η Virgo; the distance between them was about 15′.

(2) January 18 (or 16), 272 BC. Mars "touched" β Scorpius. Actually, the distance between the star and the planet was about 50′ on January 18 and about 15′ on January 16.

(3) September 4, 241 BC. Jupiter "covered" δ Cancer; the calculation shows that the actual distance was not less than 25′.

(4) March 1, 229 BC. Saturn was within "two units (digits)" from γ Virgo. Clearly, the authenticity of this observation depends on the particular meaning of the term *digit*.

It is obvious that this cannot be treated as a solution of the problem. In fact, the solution is not based on a comparison of the observations with true astronomic data (calculated), and not even on Ptolemy's intervals of time between successive events, but on interpretations of the names of months given by Ptolemy, and on astronomic data (such as longitudes of the sun and the planets) that were *calculated* by Ptolemy with the help of his theory and added to the descriptions of the events (and the fact that they had been calculated he indicated in the text of the *Almagest*). As for the authentic ancient data about the events that Ptolemy *quoted* and that were not a result of his calculations, they were completely ignored by the chronologists. These data are the years of the coverings and the very facts of occurrences of the coverings.

Note that the medieval solution agrees perfectly with the date for the star catalog of the *Almagest* obtained in Chapter 7 from our statistical analysis of latitudes. If we assume that the *Almagest* is an integral text (as most historians do), we must accept the medieval solution as the true one. However, we cannot merely ignore the existence of the second (ancient) solution, which is 1200 years to the past from the medieval one; its existence implies another hypothesis. Note that this solution does not agree with the traditional one, as well as with the traditional date for the *Almagest*. The existence of this solution may be explained by various reasons, for example, by a periodicity in coverings of stars by planets. Indeed, the configuration of planets of the solar system varies periodically with time; the coverings are determined by this configuration, so they must repeat periodically, and there is no wonder that a second solution exists (see Figure 10.1).

Corollary. *The medieval solution of the dating problem in Assertion 1 implies that the Nabonassar era (in the chronology of the Almagest) started in 480–490 AD.*

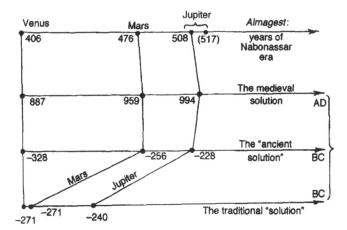

Figure 10.1. Positions in the time axis of coverings of stars by planets indicated in the *Almagest* (in Ptolemy's chronology). For comparison, computed moments of the events are shown (in two axes in the middle), as well as traditional dates. It is obvious that the traditionally accepted dates do not match Ptolemy's description.

3. A date from lunar eclipses

Twenty-one lunar eclipses mentioned in the *Almagest* were observed by different astronomers between the years 26 and 881 of the Nabonassar era. Ptolemy listed the following characteristics of the eclipses:

(1) The year of the eclipse, in terms of a chronological era given in the ancient document used by Ptolemy. Usually, this is followed by the corresponding year of the era of Nabonassar, calculated by Ptolemy. In the few cases when it is not, it is easy to carry out the recalculation using the relations between various eras, derivable from the available data.

(2) The phase of the eclipse as indicated in the ancient document. Recall that the *Almagest* contains a theory of lunar motion, but this theory does not enable Ptolemy to calculate phases of the eclipses. This is why Ptolemy quoted the phases without comments of his own. A theory of lunar motion so advanced as to calculate phases of eclipses only appeared as late as in the 19th century.

(3) The date of the eclipse and the time of "the middle of the eclipse". These data are results of Ptolemy's calculations (which he states), so they are of no interest for our dating.

Table 10.1.

No. of the eclipse	Year of the era of Nabonassar	Hour of the midtime of the eclipse in Alexandria (Ptolemy's calculation)	Phase of the eclipse (in standard units)
1	26	21	total
2	27	23	3
3	27	20	6
4	127	5	3
5	225	22	6
6	246	24	3
7	256	23	2
8	366	6	1
9	367	23	total
10	546	19	9
11	547	1	total
12	547	2	total
13	574	2	7
14	607	22	3
15	870	20	2
16	878	23	total
17	880	22	10
18	881	4	6

(4) The place where the eclipse had been observed. However, most lunar eclipses are observable from any place in the hemisphere, so these data are of little importance for us.

Thus, only the data (1) and (2) are really important for our dating problem. So we use the year of the eclipse (in terms of the era whose starting year we assume to be unknown, but which we will find from the solution of the problem), and its phase.

Recall that the phase of an eclipse is the maximal part of the moon's diameter which is shaded; the phase is measured in the units equal to 1/12 of the diameter. For total eclipses (the ones when the moon is completely covered by the earth's shade), the phase is the width of the shade crossed by the moon. For three eclipses (of 21) Ptolemy indicates no phase. But one can observe at least one lunar eclipse a year from any point of the earth's surface, so indication of a lunar eclipse without a phase provides no information. Therefore we exclude these three eclipses from further consideration. The remaining 18 eclipses are listed in Table 10.1.

(1) The problem of dating lunar eclipses in the *Almagest* may be stated as follows. We need to find a set of 18 lunar eclipses to satisfy the following conditions: Each eclipse must have the phase as fixed in the *Almagest* (to an accuracy within one unit). The phases of eclipses were

observed by medieval astronomers with a sufficient accuracy, and they were not recalculated. Therefore we assume that the phases of the eclipses are given to an accuracy within one unit (because the phases are given in integral of units).

(2) The intervals between consequent eclipses must agree with the ones given in the *Almagest*. However, Ptolemy used various documents, in which different eclipses are attributed to the years related to different eras. We should not expect an accuracy below two years (between any two subsequent eclipses). The reason is that different eras fix different months for the start of a year, so a recalculation from one era to another may introduce errors up to a year. The resulting error in the difference of two dates may amount to two years.

We have solved this problem with the help of computer calculations (from the modern theory of lunar motion), and tested the method by comparison with the well-known Kanons of the eclipses[50,41]. We considered *all* eclipses in the interval 900 BC–1600 AD. The results we have obtained may be formulated as follows.

Assertion 2. *The above problem has the unique solution that satisfies the condition 2) to an accuracy within 3 years. The corresponding set of eclipses is given in Table 10.2; all the eclipses are medieval.*

The solution is stable with respect to variations of intervals between subsequent eclipses. Ptolemy used many documents that described the eclipses, and the documents used various chronological eras. For example, the eclipses with the numbers 1–3 were dated in the documents (according to Ptolemy) in the years of the era of Mardokempad, the eclipses 4 and 5 in the era of Nabonassar, the eclipses 6 and 7 in the era of Darius, 8 and 9 in the dates of Athenian magistracy, 10, 11 and 12 in the dates of the third Kallippic period, 13 in the era of Philomotor, and 15–18 in the era of Hadrian.

As we have seen, Ptolemy's recalculations from one era to another introduced errors, sometimes amounting to ten years. This means that Ptolemy was not always aware of the exact starting dates of the eras. Therefore, we regarded temporal intervals between the dates expressed in terms of the same era as more reliable.

This is why we proceeded with our calculations, in order to see whether any other solutions appear if we increase the level of admissible errors in the intervals. We left the accuracy of 3 years for the intervals between the eclipses dated in the same eras and 30 years for the rest. It is interesting that the eclipses that are dated in terms of a particular era form a compact group in the time axis, i.e., they lie within small temporal intervals, while the intervals between successive eclipses dated in terms of different eras are separated by decades and centuries. Apparently, each group is the result of a homogeneous

Table 10.2.

No. of the eclipse	Year, AD	Month	Day	Hour (Greenwich)	Phase	Coordinates of zenith at the point of the eclipse on the earth's surface	
						longitude	latitude
1	491	8	5	16	11.1	110	−17
	or						
	492	1	30	6	16.7	123	17
2	494	6	5	1	2.0	−28	−22
3	496	11	6	21	5.0	27	17
4	594	8	6	23	4.0	16	−17
5	693	3	27	14	5.6	138	−4
6	717	6	28	13	3.0	155	−23
7	728	5	27	21	2.5	31	−22
8	840	5	20	5	1.4	−77	−21
9	843	3	19	19	14.1	73	−1
10	1019	9	16	23	9.4	10	−1
11	1020	3	12	7	18.1	−111	1
12	1020	9	4	23	18.7	13	−6
13	1046	4	23	7	6.6	−116	−14
14	1079	1	20	3	4.0	−48	19
15	1344	9	23	1	2.4	−31	3
16	1349	6	30	23	21.7	1	−23
17	1349	12	25	12	9.8	178	23
18	1350	6	20	17	5.8	103 ·	−23

group of observations, made (according to Ptolemy) by representatives of the same astronomic school, and probably more or less in the same place. So, it is natural to think that the intervals between eclipses within each group are more accurate than the intervals between groups, obtained from the latest chronological work and recalculations.

Assertion 3. *The solution in Assertion 2 is the only solution of the eclipses dating problem whose intervals between successive eclipses deviate from the ones that follow from the dates of the Almagest by at most 3 years for the eclipses dated in terms of the same era and by at most by 30 years for the eclipses dated in terms of different eras.*

If we admit a 4 year error for all intervals, a new "solution" emerges, in which all intervals but one differ from the ones ensuing from the text of the *Almagest* by less than four years, and this one differs by eight years. In principle, this single deviation may be treated as an outlie, or explained by omitting an intermediate eclipse (actually, Ptolemy mentions an eclipse, but does not supply any description of its parameters). This "solution" is close to the traditional one, but does not coincide completely with traditional dates. Figure 10.2 displays two histograms that demonstrate distribution of deviations

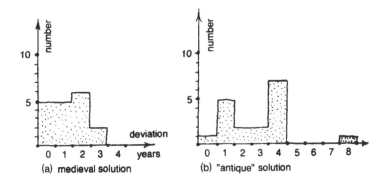

Figure 10.2. Comparison of the "antique" and the "medieval" solutions. It is obvious that the "antique" solution matches much worse.

of the intervals between the eclipses from the ones that ensue from the description in the *Almagest*, for the two solutions. It is obvious that the first (medieval) solution matches much better than the second (ancient) one.

Like in the case of star coverings, we have a periodicity of lunar eclipses, and the existence of the second solution is due to the approximate periodicity in the evolution of the configuration formed by the sun, the earth and the moon. The period is several centuries long, but the periodicity is not quite exact, so the second solution does not match exactly with the first.

4. Chronology of the *Almagest*

Our dates for coverings of stars lead to a date for the beginning of the Nabonassar era in 470–490 AD. More exactly, because of inconsistencies in the text of the *Almagest*, we get four dates, 477 AD, 481 AD, 483 AD, and 486 AD. As for dating from lunar eclipses, it leads to the date 465 AD for the beginning of the era of Nabonassar. How can we estimate the accuracy of this date? A comparison of the intervals between the dates of the eclipses given in the *Almagest* with the true intervals (between the calculated dates of the eclipses in our solution) shows that the global chronology of the *Almagest* is not free of errors; the errors in the case of the eclipses are similar to that in the case of coverings, up to about eleven years. Thus, the accuracy of relations between the dates of various eras is approximately 10–15 years.

The dates we obtain from coverings and from lunar eclipses agree perfectly; both imply the interval 460–490 AD for the beginning of the era of Nabonassar.

Now we have a possibility to reconstruct the chronological scale used in the *Almagest*. Ptolemy mentions the following historic events related to Assyria,

Egypt and Rome:

   0) the reign of Darius,

   1) the reign of Philadelphus,

   2) the beginning of the Kallippic period,

   3) the death of Alexander (it is usually understood that 4) Ptolemy means Alexander of Macedon, but in fact he simply wrote "Alexander"),

   5) the beginning of the Chaldean era,

   6) the beginning of the era of Dionysius,

   7) the reign of Augustus (book III.7),

   8) the reign of Domitian (book VII.7),

   9) the reign of Trajan (book VII.7),

   10) the reign of Hadrian,

   11) the reign of Antoninus.

Our dating implies the dates for all these events (to an accuracy within five years):

0) the beginning of the era of Nabonassar:	460–490 AD,
1) the reign of Darius:	685–715 AD,
2) the reign of Philadelphus:	840–855 AD,
3) the beginning of the Kallippic period:	875–910 AD,
4) the death of Alexander:	885–915 AD,
5) the beginning of the Chaldean era:	900–935 AD,
6) the beginning of the era of Dionysius:	915–945 AD,
7) the reign of Augustus:	1175–1205 AD,
8) the reign of Domitian:	1290–1320 AD,
9) the reign of Trajan:	1310–1340 AD,
10) the reign of Hadrian:	1310–1345 AD,
11) the reign of Antoninus:	1330–1345 AD.

Conclusions

1) The ensuing chronology agrees perfectly with the date for the star catalog of the *Almagest* we obtained in Chapter 7 (the 10th century AD). According to this chronology, the following events are to be attributed to the 9th and 10th centuries:

- the observations of coverings of stars,
- a majority of the observations of lunar eclipses,
- starting points of most important chronological eras, such as the era of Philadelphus, Kallippic periods, the era of Alexander, Chaldean era, the era of Dionysius; total, five eras of eleven mentioned in the *Almagest*.

2) The dating interval for the *death of Alexander*, 885–915 agrees with the reign of the *unique* emperor in Byzantine history (and in the history of all Western Europe) whose name was Alexander, 912–913 AD.

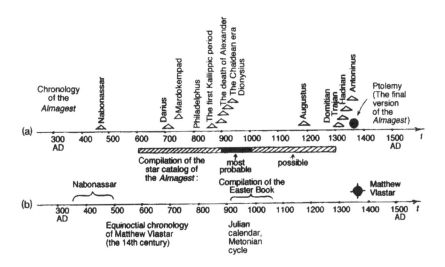

Figure 10.3. The restored chronology of the *Almagest*, compared with the chronology of a well-known ecclesiastic writer of the 14th century, Matthew Vlastar. The two chronologies agree well.

3) The dating interval for the epoch of the Kallippic era covers 877 AD, the first year of a Great Indiction. Recall that each Great Indiction is 532 years long; the Great Indiction is the period of medieval calendar, after which the dates (days of months) of all basic calendar events (such as indicts, lunar and solar cycles, etc.) repeat. Also, a shorter period was in use, called the *Kallippic period* (cycle), 76 years long. The length of a Great Indiction is exactly seven times the length of a Kallippic period, so it is natural to conjecture that a Kallippic period is just a subdivision of the Great Indiction, and so the beginning of the Great Indiction must coincide with the beginning of the first Kallippic period. This conjecture is confirmed by the above chronological scale: the first Kallippic period begins in 877 AD, exactly where the Great Indiction starts.

Appendix 1. Formulas for planetary positions

Here we give formulas we used for determination of planetary ecliptic coordinates in the past. In the formulas, t stands for the time counted as in the previous chapters, that is, from January 1 1900, 12:00 ephemerid time to the past in Julian centuries (see Chapter 1; Julian century is 36525 ephemerid days); L is the mean longitude of the planet at the moment t, π is the mean longitude of the perigee of the orbit of the planet, e is the eccentricity of the orbit, i is the angle between the plane of the planet's orbit and the ecliptic,

Ω is the longitude of the ascending knot of the orbit, and a is the principal semiaxis of the orbit. The coordinates L, π and Ω are referred to the ecliptic at the moment t.

EARTH

$L = 99°41'48''04 + 1296027768''13t + 1''089t^2,$
$\pi = 101°13'15''0 + 6189''03t + 0''012t^2,$
$e = 0.01675104 - 0.0004180t - 0.000000126t^2,$
$i = 0,$
$a = 1.00000023.$

MARS

$L = 293°44'51''46 + 68910103''83t + 1''1184t^2,$
$\pi = 334°13'05''53 + 6626''73t + 0''4675t^2 - 0''0043t^3,$
$\Omega = 48°47'11''19 + 2775''57t - 0''005t^2 - 0''0192t^3,$
$e = 0.09331290 + 0.000092064t - 0.000000077t^2,$
$i = 1°51'01''20 - 2''430t + 0''0454t^2,$
$a = 1.52368840.$

VENUS

$L = 342°46'01''39 + 210669162''88t + 1''1148t^2,$
$\pi = 13009'49''8 + 5068''99t - 3''515t^2,$
$\Omega = 75°46'46''73 + 3239''46t + 1''476t^2,$
$i = 3°23'37''07 + 3''621t - 0''0035t^2$
$e = 0.00682069 - 0.00004774t + 0.000000091t^2,$
$a = 0.72333162.$

JUPITER

$L = 238°02'57''32 + 10930687''148t + 1''20486t^2 - 0.005936t^3,$
$\pi = 12°43'15''34 + 5795''862t + 3.80258t^2 - 0''91236t^3,$
$\Omega = 99°6'36''19 + 3637''908t + 1''2680t^2 - 0''03064t^3,$
$e = 004833475 + 0.000164180t - 0.0000004676t^2 - 0.0000000017t^3,$
$i = 1°18'31''45 - 20''506t + 0.014t^2,$
$a = 9.554747.$

SATURN

$L = 266°33'51''76 + 4404635''5810t + 1''16835t^2 - 0''021t^3,$
$\pi = 91°05'53''38 + 7050''297t + 2''9749t^2 + 0''0166t^3,$
$\Omega = 112°47'25''40 + 3143''5025t - 0''54785t^2 - 0.0191t^3,$
$e = 0.058932 - 0.00034550t - 0.000000728t^2 + 0.00000000074t^3,$
$i = 2°29'33''07 - 14''108t - 0''05576t^2 + 0''00016t^3,$
$a = 9.554747.$

These relations determine the positions of the earth and the planets at the moment t; we used them to determine the direction of the line through the earth and the planet and compare it with the direction towards the star supposed to be covered. An analysis of stability of the calculation shows that its accuracy is 1′, which is sufficient for our purposes.

Appendix 2. Lunar eclipses in the *Almagest*

1) "Then of the three ancient eclipses observed in Babylon, of which we spoke, the first is recorded as having taken place in the year 1 of Mardokempad ... And the eclipse began, it is started, more than one hour after the rise of the moon, and the eclipse was total" (Ref. 24, p. 123, book IV.6).

2) "The second of the eclipses is recorded as having occurred in the year 2 of Mardokempad ... And there was an eclipse, it says, of 3 digits from the southern end at midnight" (Ref. 24, p. 123, book IV.6).

3) "The third of the eclipses is recorded as having taken place in the same year 2 of Mardokempad ... And the eclipse began, it says, after the rise of the moon, and there was an eclipse of more than half from the northern end" (Ref. 24, p. 123, book IV.6).

4) "For the year 5 of Nabopolassar (which is the year 127 of Nabonassar ...) the moon began to be eclipsed in Babylon; and the greatest extent of the eclipse was 1/4 of the diameter from the south" (Ref. 24, p. 172, book V.14).

5) "Again, in the year 7 of Cambyses (which is the year 225 of Nabonassar ...) the moon was eclipsed to the extent of a half of its diameter from the north" (Ref. 24, p. 172, book V.14).

6) "The second is the one Hipparchus used, occurring in the year 20 of Darius, successor of Cambyses ... And here likewise the moon was eclipsed to a breadth of 2 digits from the southern side" (Ref. 24, p. 137, book IV.9).

7) "We then took, first, the eclipse observed in Babylon in the year 31 of Darius ... and the moon was eclipsed to a breadth of 2 digits from the southern side" (Ref. 24, p. 136, book IV.9).

8) "Now he says there three eclipses were given out by those crossing over from Babylon as having been observed there, that the first of them occurred in the Athenian magistracy of Phanostratus ... the moon was eclipsed to the extent of a small bit of its circle ... And he says it was still eclipsed when setting" (Ref. 24, p. 140, book IV.11).

9) "Again, he says, the next eclipse occurred in the Athenian magistracy of Phanostratus ... and the moon was eclipsed ... This date is the year 366 of Nabonassar" (Ref. 24, p. 140, book IV.11; no phase indicated).

10) "And he says the third eclipse occurred in the Athenian magistracy of Evandrus ... And the eclipse, he says, was total ... This date is the year 367 of Nabonassar" (Ref. 24, p. 141, book IV.11).

11) "And next we shall pass to the three later eclipses set out by him, which he says were observed in Alexandria. He says the first of these occurred in the

year 54 of the Second Kallippic Period ... the moon began to be eclipsed 1/2 our before rising and returned to its full size in the middle of the third hour" (Ref. 24, p. 141, book IV.11).

12) "The next eclipse occurred, he says, in the year 55 of the same period ... and the eclipse was total" (Ref. 24, p. 142, book IV.11).

13) "And he says the third eclipse occurred in the same year 55 of this Second Period ... and the eclipse was total" (Ref. 24, p. 142, book IV.11).

14) "In the year 7 of Philometor then (i.e., the year 574 of Nabonassar ...), in Alexandria the moon was eclipsed up to 7 digits from the north" (Ref. 24, p. 196, book VI.4).

15) "For example, from the observation of the eclipse in the year 32 of the Third Kallippic Period ... " (Ref. 24, p. 80, book III.1; no phase indicated).

16) "Once again, in the year 37 of the Third Kallippic Period (which is the year 607 of Nabonassar...), the moon began to be eclipsed ... and was obscured at the most 3 digits from the south" (Ref. 24 p. 196, book VI.4).

17) " ... From the eclipse in the year 43 of the same period ... " (Ref. 24, p. 80, book III.1; no phase indicated).

18) "Second, we took that observed in Alexandria in the year 9 of Hadrian ... and the moon was eclipsed likewise to the extent of 1/6 of its diameter from the southern side" (Ref. 24, p. 136, book IV.9).

19) "Again, of the three eclipses we have chosen from those most carefully observed by us in Alexandria, the first occurred in the year 17 of Hadrian ... And the eclipse was total" (Ref. 24, p. 129, book IV.6).

20) "The second occurred in the year 19 of Hadrian ... And there was an eclipse to the extent of 1/2 + 1/3 of the diameter from the northern side" (Ref. 24, p. 129, book IV.6).

21) "The third of the eclipses occurred in the year 20 of Hadrian ... And there was an eclipse to the extent of 1/2 of the diameter from the northern side" (Ref. 24, p. 129, book IV.6).

Part Three
Addendum

Addendum

Introduction. Problems and hypotheses connected with dating the *Almagest*

> "*Rhodes* is distorted *Rhoda*, the name of town in Türingia where, according to the folk book, Faustus was born" (Ref. 42, p. 408).

Chapter 10 ends the main text of this book, and we could stop our exposition here, leaving it to the reader to cope with our unexpected change of date for the *Almagest* that follows from our results. However, after a discussion between the authors, we decided that this would not be quite proper. Of course, the idea of displacing an important literary monument of the history of astronomy from the 2nd century AD to the Middle Ages must be somewhat difficult for an unprepared reader; this displacement, if considered separately from the knowledge and experience that our readers undoubtedly have, could naturally raise a protest. Therefore we include this Addendum. We are not going to "help" the reader by forcing our point of view; to the contrary, we are going to raise still more problems. We hope thus to raise interest in a study of problems of chronology. On the other hand, we do not conceal our point of view, so we will formulate some conjectures.

Together with some irrefutable facts, this Addendum suggests several conjectures and unsolved problems connected with the position of the *Almagest*

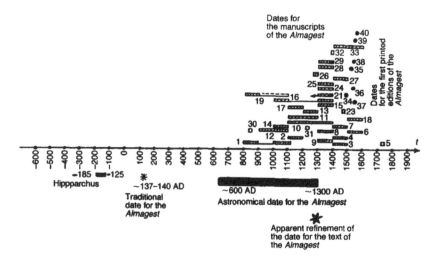

Figure A1. Dates for manuscripts of the *Almagest*.

in the history of science, in particular, with the date of its creation. *This material is additional to the main body of the book.* A reader only interested in astronomic and statistical aspects of the problem will lose nothing if this Addendum is skipped. To the contrary, a reader wishing to extend his/her knowledge about the situation with the problem of dating the *Almagest* will hopefully find new facts of interest.

1. On dates for various manuscripts and printings of the *Almagest*, and how they agree with the date from astronomic analysis

In this section we compare the date for the *Almagest* we have inferred from our astronomic analysis with the dates of extant manuscripts of the *Almagest* as known in traditional chronology; we also provide information about the dates of the first printed editions of the *Almagest*. The data we use are from Ref. 22, which contains a complete list of the most ancient Greek, Latin and Arabic manuscripts of the *Almagest*. For reader's convenience, we give the chronological diagram (Figure A1) where the dates for the texts are plotted along the horizontal times axis; we also mark the interval of admissible dates for the catalog of the *Almagest* obtained from our statistical analysis of the contents of the star catalog. We also give the life dates of several medieval persons who were connected with astronomy, with finding ancient manuscripts and with establishing the chronology (Figure A2).

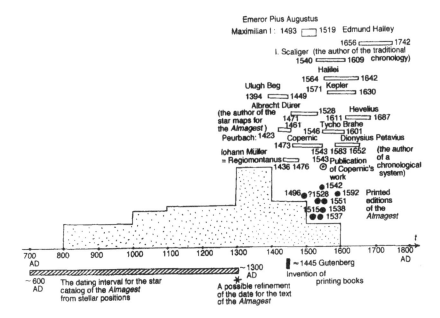

Figure A2. Distribution of traditional dates for manuscripts of the *Almagest*. The years of lives of some historical persons connected with the development of astronomy are indicated.

1. Greek manuscripts of the Almagest.

1. *Paris Codex 2389.* This manuscript and no. 19 below are believed to be the oldest manuscripts of the *Almagest* (see Ref. 22, p. 19 and further). It is supposed that the manuscript was originally in Florence, whence Catherine de Medici probably brought it to Paris; on her death it came to a library (now called Bibliothèque Nationale). *It bears the stamp in gold of Henri IV.* Opinions about the date of creation of this copy vary. We should stress a general feature: dating manuscripts of the *Almagest* is very difficult unless they tell the date themselves. In this case, the stamp of Henri IV could serve as an indication; *this attributes the manuscript to the 11th century or to the beginning of the 12th century.* Nowadays, the method of *paleography* is sometimes used for dating manuscripts which is based on an analysis of the shape of letters. Not going into details about this method, let us only note its ambiguity. On the basis of this method, Halma suggested to attribute the manuscript to the 7th or to the 8th century, but in the traditional history of astronomy, attribution to the 11th century (also paleographic) is generally accepted; see a discussion of this dating problem in Ref. 22, p. 19. In our diagram, we mark both dates, in the 9th century (the *paleographic hypothesis*) and the 11th–12th century (the stamp of Henri IV).

Passing to other manuscripts, we should note that unfortunately, the principles of attributing manuscripts to particular centuries are practically out of
discussion in Ref. 22. Most information about attributing dates involve paleographic analysis; so we will only indicate formally the supposed dates of
creation of manuscripts generally accepted in the traditional history of science. Most of the dates are supplemented in Ref. 22 by a note "about", which
emphasizes difficulties in dating.

2. *Paris Codex 2390.* Attributed to *about the 12th century.*

3. *Paris Codex 2391. About the 15th century.*

4. *Paris Codex 2392. About the 15th century.* The text is incomplete, a very
bad copy.

5. *Paris Codex 2394.* A copy, made in 1733.

6. *Vienna Codex 14. About the 15th century.*

7. *Venice Codex 302. About the 15th century.*

8. *Venice Codex 303. About the 14th century.*

9. *Venice Codex 310. About the 14th century.*

10. *Venice Codex 311.* In the catalog of Zanetti, *attributed to about the
12th century,* but in Peters' opinion, *the true date is much later.* In Morelli's
opinion, the manuscript is a later copy of *Venice Codex 313* (dated about the
10th or the 11th century) or even of *Venice Codex 303* (attributed to about the
14th century). Here again we see that the dates for the manuscripts are far
from being quite definite.

Summing up all the opinions, we get the dating interval from the 12th to
the 14th centuries.

11. *Venice Codex 312.* Zanetti dates this manuscript *about the 12th century,*
and Morelli *about the 13th century.*

12. *Venice Codex 313.* Zanetti attributes this to *about the 10th century,*
and Morelli to *about the 11th century.*

13. *Laurentian Codex. Pluteus 28, 1. About the 13th century.*

14. *Laurentian Codex. Pluteus 28, 39. About the 11th century;* only contains Books VII and VIII.

15. *Laurentian Codex. Pluteus 28, 47. About the 14th century.*

16. *Laurentian Codex. Pluteus 89, 48. About the 11th century.* A very well
written manuscript, but has very much in common with *Venice Codex 310,*
attributed to the 14th century.

17. *Vatican Codex 1038. About the 12th century.*

18. *Vatican Codex 1046. About the 16th century.*

19. *Vatican Codex 1594. About the 9th century.* The best of the Greek
manuscripts of the *Almagest.* Unfortunately, Ref. 22 describes no base for
the date, but it is noted that this manuscript has much in common with *Venice
Codex 313* "indicating a common origin"(Ref. 22, p. 21). But *Venice Codex
313* (no. 12 above) is dated *the 10th or the 11th century.*

20. *Vatican Codex, Reg. 90.* Peters and Knobel remark that "This codex

is probably not very old ... " (Ref. 22, p. 21), but give no date. We omit it in our diagram.

21. *Bodleian Codex 3374. Before the 14th century.* A modern copy, very well written, without variants.

2. Latin manuscripts of the Almagest

22. *Vienna Codex 24 (Trapezuntius).* A perfect codex, titled *Magnæ compositionis Claudii Ptolomæi libri a Georgio Trapezuntio traducti.* The date is given in the end: "Finis 17 Marcii, 1467".

23. *Laurentian Codex 6. Dated in the interval 1471–1484.* Supposed to be a translation from Greek. Neatly and accurately written.

24. *Laurentian Codex 45. About the 14th century.* A perfectly written manuscript, has many variants. This and the three following manuscripts are believed to be copies of a translation from Arabic.

25. *British Museum Codex. Burney 275. About the 14th century.* Believed to be a translation from Arabic. A very good copy of the *Almagest*, perfectly written.

26. *British Museum Codex. Sloane 2795.* The manuscript is considered to be a translation from Arabic; the date according to Thompson, is *about the 13th century, probably earlier but not before 1272.* Beautifully written, but contains many errors.

27. *Crawford Codex. About the 15th century.* A very good manuscript, considered to be a translation from Arabic.

28. *New College, Oxford, No. 281.* A very imperfect copy of the translation of Gerard of Cremona, hence created not before *the 14th century.*

29. *All Souls College, Oxford, No. 95.* Also the translation of Gerard of Cremona, with several books omitted. *Not before the 14th century.*

3. Arabic manuscripts of the Almagest

30. *Codex Laurentianus 156.* A carefully written manuscript. Presumably a copy of the translation made by Al Mamon *about 827 AD.*

31. *British Museum 7475.* An incomplete copy of the *Almagest*. Dated 615 after Hejira, hence, after the generally accepted recalculation into the years AD, 1218. Many longitudes and latitudes differ from that given in the other manuscripts.

32. *Bodleian Arabic Almagest, Pocock 369. Dated 799 year after Hejira, that is, 1396 AD.* A well-written copy.

33. *British Museum Arabic Manuscripts, Reg. 16, A. VIII. About the 15th or the 16 century.*

In our chronological diagram we depict the estimates for the dates of the above manuscripts as white intervals; for example, the interval with the ends

1272 and 1300 is the dating interval for no. 26. If only the dating century is known, we depict an interval that covers the century.

Now we pass to the *first printed editions of the Almagest*. In order not to confuse them with the dates for the manuscripts, we depict them in Figure A1 by black points.

34. The first printed edition of the *Almagest* is supposedly, 1496. *Joannis de Monte Regioet Georgii Purbacho Epitome in Cl. Ptolemaei magnum compositionem (Venice, 1496?).*

We also give here some information about the first editions of the *Almagest* collected by N. A. Morozov in the book storage of the Pulkov observatory[4]. Concerning this book, he remarked, "There is, for example, the printed book of Ioann Regiomontanus and Georgius Peurbach *Epitome of the Great Composition of Claudius Ptolemy* on which it is indicated, if only what I know about it is true, Venice, 1496" (Ref. 4, vol. 4, pp. 218–219). According to the information received by the authors, the book only contains the text of the *Almagest*, tables missing, so it contains no star catalog. See also Ref. 4, vol. 4, pp. 195–196.

35. 1515 edition. *Almagestu Cl. Ptolemaei Phelusiensis Alexandrini. Anno Virginei Partus 1515* (edited by Liechtenstein). See Ref. 4, vol. 4, pp. 195–196.

This Latin edition was published by Liechtenstein in 1515 in Venice. Baily thinks[28] that this is a translation from Arabic, while the edition of 1537 he considers to be a translation from Greek. *The 1515 edition is very rare.* According to Baily, Lalande saw this book, the unique copy of which was in the library of the London Royal Astronomic Society. N. A. Morozov communicates that the book also was in the book storage of the Pulkov observatory.

36. 1528 edition. In the book storage of the Pulkov observatory there is the book *Claudii Ptolemaei Phelusiensis Alexandrini. Anno Salutis, 1528, Venice, translation of Trapezuntius.* We studied the star catalog in this edition in the same way we did the one in the book of Peters and Knobel[22]. *The result of the analysis of the catalog in this edition coincides with the one obtained from the analysis of Ref. 22.*

The next two, the Latin 1537 (Cologne) and the Greek 1538 (Basel) editions are the most famous.

37. The Latin 1537 edition. *Cl. Ptolemaei. Phelusiensis philosophi et matematici excellentissimi Phaenomena, stellarum MXXII. fixarum ad hanc aetatem reducta, atque seorsum in studiorum gratiam. Nunc primum edita, Interprete Georgio Trapezuntio. Adiecta est isagoge Ioannis Noviomagi ad stellarum inerrantium longitudines ac latitudines, cui etiam accessere Imagines sphaerae barbaricae duodequinquaginta Alberti Dureri. Excessum Coloniae Agrippinae* (it is now accepted that this is Cologne—*Authors*). *Anno D.M.XXXVII, octavo Calendas 5 (?—Authors) Septembres.*

38. Greek 1538 edition. *Κλ. Πτολεμαιου Μεγαλης Συνταξεως Βιβλ Π. Θεωνο Αλεξανδρεως εις τα αυτα υπομνηατον Βιβλ. IA. Claudii Ptolemaei Magnae Constructionis, id est perfectae coelestium motuum pertractationis*

Lib. XIII. Theonis Alexandrini in eosdem Commentariorum Libri XI. Basileae apud Ioannem Walderum An 1538. C. puv. Caes. ad Quinquennium.

39. The second Latin translation, 1542 (see Ref. 24, vol. 4, pp. 195–196).

40. The third Latin translation, 1551 (see Ref. 4, pp. 195–196).

41. *Clavdii Ptolemaei inerrantium stellarum Apparitiones, et significationum, collectio. Federico Bonaventura interprete. Urbini 1592.*

Let us now mark in the chronological diagram the interval 600 AD–1300 AD where, by the above analysis, the true date of creation of the *Almagest* must be. *It is obvious that the interval agrees perfectly with the set of all dates of the extant manuscripts and the first printed editions of the Almagest.* The abundance of the manuscripts itself may indicate that the *Almagest* had been created at that time and began to spread at once, being an important scientific treatise, treated not as a literary monument, but as a scientific handbook, the methods of which found immediate applications to concrete problems in astronomy, navigation, etc. *This agreement of our date from the astronomic analysis with the dates of the extant manuscripts of the Almagest appears to be no mere occasion.* Our conjecture is that the *Almagest* was not left unattended for hundreds of years (in the traditional version, from the 1st century AD until the Renaissance), but was at once after its creation involved in scientific use, commented, copied, and soon, in the beginning of the 16th century, printed. Note that *handwritten books still existed after the invention of printing*; many decades, the copyists coexisted with printing; sometimes they even produced handwritten copies of printed editions. The explanation of this is quite simple: copying manuscripts in hand was still cheaper than printing. *Therefore, some manuscripts of the Almagest now considered to be ancient (that is, created before the invention of printing, in the interval 10th–15th centuries) could as well have appeared after the first printed books.*

We will give brief information that demonstrates that *handwritten books still existed long after the start of printing* (see Ref. 43 for details). For example, in the Balkans, as late as in the 19th century "handwritten books competed successfully with printing" (Ref. 43, p. 26). With few exceptions, the Irish literature of the 7th–17th centuries is all handwritten (Ref. 43, p. 28). In the library of John Dee, English mathematician and astrologer of the 16th century, 300 books of 400 were handwritten (Ref. 43, p. 56). Until 1500, 77% of all printed books were in Latin, because Latin fonts were easy to manufacture. *Prints of other languages came into use very slowly*, because of difficulties in manufacturing accents and other specific symbols. So *centuries after the start of printing*, "copyists of Greek, Arabic and Hebrew books were beyond competition ... " (Ref. 43, p. 57). Especially many handwritten books existed in Greece. "Because of absence of printing shops, in Greece books were copied in hand ... " (Ref. 43, p. 106). "Luxurious Greek codices with texts of antique authors were manufactured to orders of humanists and collectors" (Ref. 43, p. 109), so what reliable results can a paleographic analysis of such manuscripts give? Giovanni Bernardo Regasola (Feliciano) created "exquisite Greek manuscripts" (Ref. 43, p. 109) in the *16th century*. Greek monasteries were especially famous for their copyists (in the printing era already). It is important to remark that *many manuscripts were copied from printed books* (Ref. 43, p. 120). By the way, such copies may be detected by comparing errors in the texts with misprints in printed editions; apparently, almost all misprints were reproduced.

Let us now return to Figures A1 and A2. Figure A2 shows the following additional data, useful for bringing in order the information connected with the *Almagest*.

Iohann Müller = Regiomontanus: 1436–1476. *Copernicus*: 1473–1543. In 1543, Copernicus' book *Revolutionibus Orbium Saelestium* was published, that pursues the scientific tradition of the *Almagest*, whose printed editions were published vigorously at the time. *Tycho Brahe*: 1546–1601. *Peurbach*: 1423–1461. *Albrecht Dürer* (the author of star maps for the first editions of the *Almagest*): 1471–1528. *Ulugh Beg*: 1394–1449. *Kepler*: 1571–1630. *Halilei*: 1564–1642. *Edmund Halley* (the presumed discoverer of the proper motion of stars). *Ian Hevelius*: 1611–1687. Roman emperor *Pius Augustus Maximilian I*: 1493–1519 (recall that according to the tradition, the *Almagest* was created in the reign of Roman emperor *Antoninus Pius Augustus*, 138–161 AD). *I. Scaliger* (the creator of the now traditional version of ancient chronology): 1540–1609; his fundamental treatise on chronology was published in 1583. *Dionysius Petavius* (a disciple of Scaliger, also one of the creators of the modern version of ancient chronology): 1583–1652. The invention of printing (Gutenberg): about 1445.

In conclusion, let us return to the problem of dating manuscripts of the *Almagest*. We have already remarked that the dates are usually from a paleographic analysis. Generally not particularly reliable, the results are still more doubtful because of the fact of common use of copying by hand after the invention of printing. Probably, some collectors ordered books to look ancient from the point of view of script. Therefore, a complete revision of dating the extant manuscripts of the *Almagest* appears to be very useful; the following questions are to be answered.

1) Where is the manuscript now (in what particular archives, museum, private collection, etc.)?

2) How was the manuscript found? To what year back from the 20th century it is possible to trace the history of the manuscript? Who found it and in what circumstances? What documental evidence hereof exists?

3) What is the date for the manuscript? Who suggested the date first and on what basis? To what extent is this date unique? Are there any other versions of the date? In mathematical terms, how many solutions does the dating problem admit?

4) Suppose the book was written in the reign of "the emperor Pius".
What particular Pius is meant? The Roman emperor of the 2nd century
AD or the Roman emperor of the 15th–16th centuries (Pius Augustus
Maximilian I)?

5) Furthermore, *most ancient names have a meaning*. For example, *Pius*
means "pious". So, the text was written in the reign of *some* pious emperor.
It should be clear that this characteristic applies to any ruler (in any country).
The absence of an unambiguous answer generates a vast arbitrariness in the
choice of the date.

In connection with item 5), the following remark seems relevant. In antiquity (up to the Middle Ages), most people had no fixed personal names in the
modern sense. The names were rather what we now call nicknames, usually
meaningful (in the language in which they were originally pronounced), and
reflected various human features. Various chroniclers (for example, living in
different towns) gave an emperor the names corresponding to his features
that were known in the particular region (in one place he was known as "pious", in another as "cruel", etc.). Pharaohs had different names before and
after coronation; since they coronated several times (to the crowns of several regions), the number of their "names" increased sharply. The father of
a Roman consul in 169 AD had 13 names, and his son 38 names; all these
names are in fact words like "strong", "lucid", etc. Roman warriors acquired
the names from the places where they had victories. B. L. Smirnov noted,
"Meaningless names are very rare … " (Ref. 44, vol. 6, p. 526). Most names
were given in maturity. "The personal name is a password, indicating the
belonging of its bearer to this or that social stratum" (Ref. 45, p. 20). "Tsar
Ivan III had the name Timofeĭ; tsar Vasiliĭ III was Gavriil … prince Dmitriĭ
was not Dmitriĭ, but Uar; one is the tsar's name, and the other is ecclesiastic"
(Ref. 46, p. 22).

Thus, even if the text indicates that it was created "in the reign of Emperor
Mighty", this provides little base for dating, because a majority of emperors
undoubtedly deserve this name.

6) Let us consider the following kind of an argument: "The astronomer A
(or the writer) A refers to Ptolemy, so A lived after Ptolemy".

This argument is quite disputable. The question is *which* Ptolemy is meant.
Not to mention the possibility of a meaningful translation of the name *Ptolemy*,
there are a lot of possibilities for identification of the Ptolemy with various
persons belonging to various times.

7) Let us consider the following argument: "the astronomer A writes that
he had read the *Almagest* of Ptolemy, so the *Almagest* had been written before
A lived". This conclusion is also disputable; it is relevant to ask: *How can
we determine what particular text of the Almagest is meant*? Can it be proved
that what is meant is the now known text of the *Almagest* (clearly of medieval
origin, as follows even from the table in Figures A1 and A2, using traditional
data only). The fact is that versions of the *Almagest* differ not only in the text,

but also in numeric data (see the examples below). Furthermore, we need a date for the life of A.

A reader should not treat the above system of questions as quibbles. To the contrary, raising these questions is the only way to find the date more or less reliably (otherwise, each date is a reflection of someone's personal point of view). Generally, it would be very useful to find the origins of all dates and to compile a table of all now accepted dates in the form "the event ... took place in ... , as ... states". Indicating each time the author of the date, we will restore the original sources of traditional dates and will make them accessible to an objective verification.

2. About R. Newton's book *The Crime of Claudius Ptolemy*

Above, we sometimes referred to a fundamental work of R. Newton[1], devoted to the *Almagest*. We should state our attitude to this book here, because various opinions about this book exist in modern astronomic and about-astronomic literature. The book exposes an investigation of the *Almagest* by astronomic, mathematical and statistical methods. It contains a lot of statistical material, obtained in long-standing activities of R. Newton, and revealing to a great extent the nature of difficulties connected with interpretation of the data contained in the *Almagest*. It should be noted that R. Newton never doubts that the *Almagest* was compiled about the 1st century AD (more exactly, somewhere in the interval 200 BC–200 AD), because, being a professional astronomer, he relies completely on the global traditional chronology, within the frames of which he treats the *Almagest*. The main conclusions of R. Newton may be summarized as follows.

1)   The data contained in the *Almagest* do not match the astronomic reality of the 1st century AD.

2)   The existing version of the *Almagest* contains not the original observational data, but a result of recalculation (that is, somebody had for some reason recalculated the data for a different epoch).

3)   The *Almagest* could not be compiled in 137 AD, where Ptolemy is traditionally attributed.

4)   Therefore, the *Almagest* was compiled in a different time, and so requires a redating (R. Newton suggests that the date should be moved to the past, to the time of Hipparchus, about the 2nd century BC).

5)   R. Newton shares the traditional assumption that it is told in the *Almagest* that the observations had been carried out by Ptolemy himself, about the beginning of the reign of Roman emperor Antoninus Pius (traditionally attributed to 138–161 AD).

Because of the obvious contradiction between the items 1–3 and 5, R. Newton concludes that Ptolemy lied (whether the *Almagest* really implies that the observations had been carried out by Ptolemy himself, about the beginning of the reign of Antoninus Pius, we will discuss below).

In other words, in R. Newton's opinion, Ptolemy (or another compiler of the *Almagest*) is a falsifier, for he passes off the result of a recalculation (of some observational data, the date of which is unknown) as the data obtained from observations of his own.

Thus, the necessity of redating the *Almagest* is proved by R. Newton, both from astronomic and statistical methods. The other question is where the date should be displaced. As we have noted, R. Newton suggests to move the date back to the epoch of Hipparchus. Other points of view are possible; we will discuss this below. In any case, R. Newton does not discuss (not even raises) the question whether any epoch displacement of the date of compilation of the *Almagest* lifts all (or most) mismatches. *As we will see, the displacement of the date to the epoch of Hipparchus does not solve the problems.* But in this case, we deal with an attempt to displace the date by as little as 220–300 years, and only to the past. The natural question arises: Is it possible to consider other displacement, also to the future, and probably by much more than 300 years? From the point of view of mathematics and astronomy, the question is quite correct, and an unbiased researcher should have to find an answer to this question.

Let us now formulate our attitude to the five conclusions of R. Newton. The first four are of astronomic and statistical nature, and since we have checked most of Newton's calculations, we have no reasons to contest them. If we assume that Ptolemy lived about the 1st century AD, the conclusion about falsification is the only one possible. But if we assume that the date of compilation of the *Almagest* differs from the traditional one (137 AD) by more than 300 years, then an epoch may exist where the *Almagest* will lift all astronomic and statistical contradictions, as well as the purported contradiction with the fact of "observations being carried out personally". Therefore we refrain from accusations for falsification brought by R. Newton. On the other hand, we must agree with R. Newton when he tells that the observational data have been recalculated for a different epoch. In conclusion of the section, we give a quotation from the preface to Newton's book (Ref. 1, p. xiii):

"This is the story of a scientific crime. By this, I do not mean a crime planned with the care and thoroughness that scientists like to think of as a characteristic of their profession, nor do I mean a crime carried out with the aid of technological gadgetry like hidden microphones and coded messages on microdotes. I mean a crime committed by a scientist against his fellow scientists and scholars, a betrayal of the ethics and integrity of the profession that has forever deprived mankind of fundamental information about an important area of astronomy and history."

And in the conclusion of his book, R. Newton writes:

"All of his own observations that Ptolemy uses in the *Syntaxis* (= the *Almagest—Authors*) are fraudulent, so far as we can test them. Many of the observations that he attributes to other astronomers are also frauds that he has committed. His work is riddled with theoretical errors and with failures of comprehension, as we saw in Section XIII.5. His models of the moon and Mercury conflict violently with elementary observation and must thus be counted as failures. His writing of the *Syntaxis* has caused us to lose much of the genuine work in Greek astronomy. Against this we can set only one possible asset, and it is questionable whether this is a contribution that Ptolemy himself made. This possible asset is the equant model used for Venus and the outer planets, and Ptolemy lessens its value considerably by his inaccurate use of it.

"It is clear that no statement made by Ptolemy can be accepted unless it is confirmed by writers who are totally independent of Ptolemy on the matters in question. *All research in either history or astronomy that has been based upon the Syntaxis must now be done again* (italics ours —*Authors.*)

"I do not know what others may think, but to me there is only one final assessment: The *Syntaxis* has done more damage to astronomy than any other work ever written, and astronomy would be better off if it had never existed.

"Thus Ptolemy is not the greatest astronomer of antiquity, but he is something still more unusual: He is the most successful fraud in the history of science" (Ref. 1, pp. 378–379).

Other scientists are also rather skeptical about the role of Ptolemy in the history of science. Thus, A. Berry tells that although in the Middle Ages the authority of Ptolemy was decisive, among modern astronomers there is much dissension about his role. It is now known (and in fact Ptolemy himself never tried to conceal this) that his works are based to a great extent on the works of Hipparchus; as for his own observations, they are mostly not very reliable, even if not faked[2].

3. Why various versions of the *Almagest* differ

It seems strange that under the same title *Ptolemy's Almagest*, medieval editors place texts that sometimes differ much (see Chapter 2). It is common to consider now that all these editions go back to the same origin (of course, lost). However, the differences and variations are far beyond copyists' errors. Sometimes the texts themselves differ, sometimes the numeric data, etc. It seems that under the title *Ptolemy's Almagest* works of a school of astronomers rather than a single work were published. Our conjecture is that the *Almagest* was not created by a single observer, but is a collectively written medieval handbook of astronomy. Possibly, the author (or the authors) collected under the same title many separate observational data, various theories, etc., that belonged to various astronomers, possibly separated by decades or even centuries. In particular, the catalog of the *Almagest* could be compiled by one observer, and the text written by other authors. Therefore we should not assume that the text of the *Almagest* and the star catalog are so inseparable. We have an impression that the text is of a survey character; it seems to explain several points of view of various specialists rather than a unitary system of views.

4. Does it follow from Ptolemy's text that he had carried out the observations himself?

A general assertion we are going to state in this section is that it does not follow at all from the Ptolemy's text that he had carried out all observations and measurements himself. The text of the *Almagest* admits many different interpretations; one of them is that the *Almagest* is not an account of an observer, but a collection of works of several astronomers, a kind of a handbook (or even a textbook) for young astronomers and generally all scientists, an exposition of various observational methods and techniques, etc. Let us look at some examples.

We use Toomer's edition of the *Almagest* [17].

Ptolemy, as he describes the meridian circle in Book I of the *Almagest*, writes, "We make a bronze ring of a suitable size (!?—*Authors*) ... we use this as a meridian circle, by dividing it into the normal 360° of a great circle, and subdividing each degree into as many parts as [the size of the instrument] allows (!?—*Authors*) ... we found an even handier way of making this kind of observation by constructing, instead of the rings, a plaque ... of stone or wood (!?—*Authors*) ... " (Ref. 17, pp. 61–62). Obviously, this is not a description of a *concrete* instrument used by Ptolemy (or somebody else) for the observations; otherwise how can we explain the ambiguities like "of a suitable size", "into as many parts as the size of the instrument allows", "a plaque of stone or wood"? All this becomes clear if we understand that we do not have an account here, but a textbook explaining to a potential student (or to a scientist) *how to construct* the necessary instruments, *how to make measurements* (in various ways), etc.

Another example is in Book VIII: " ... to the beginning [of the reign] of Antoninus, which was when we made the majority of our observations of the positions of the fixed stars" (Ref. 17, p. 328). In traditional history of astronomy, this phrase is understood as the statement that Ptolemy had made the observations himself about the beginning of the reign of Roman emperor Antoninus Pius (in traditional chronology, 138–161 AD). But the reader can see that the phrase admits different interpretations. First of all, who is "we" who had made observations? Was it Ptolemy himself, or his predecessors belonging to the same school? Furthermore, what means "the majority of observations"? Apparently, the words "we measured" are to be attributed to Ptolemy's specific style, but not to his personal participation in the measurements.

Let us now look at how Ptolemy introduces to the star catalog. It would be natural to expect that the observer should describe clearly the procedure he had used, the reference stars, etc. *However, Ptolemy writes nothing like that.* Here is his text:

"Hence, again using the same instrument [as we did for the moon] (astrolabon — *Authors*), we observed as many stars as we could sight down to the sixth magnitude. [We proceeded as follows.] We always arranged the first of the above-mentioned astrolabe rings ... [to sight] one of the bright stars whose position we had previously determined by means of the moon, ... "(Ref. 17, p. 339). A description of the technique of measuring stellar coordinates follows, in which the latitude is measured in reference to bright stars and the latitude in reference to the ecliptic ring of the astrolabe. This description is also given *in very general terms*. Then follows the phrase "In order to display the arrangement of stars on the solid globe according to the above method, we have set it out below in the form of a table in four sections" (Ref. 17, p. 399). Further, the notation used in the tables is explained. The "table" is exactly the famous star catalog. Thus, we see that Ptolemy only needs a catalog for the description of how to make a globe. Again, this resembles an instructive text. By the way, as he describes the table, Ptolemy again mentions Antoninus: " ... In the second section, its position in longitude, as derived from observation (whose, is not indicated — *Authors*), for the beginning of the reign of Antoninus ... "(Ref. 17, pp. 339–340). Again, this phrase does not state that the observations had been carried out by the author. The phrase may be understood as the statement that the author reduces the catalog to the beginning of the reign of Antoninus (by the way, the *Almagest* indicates no dates of his reign). As we know, reduction of a catalog in ecliptic coordinates to a given date is *elementary*; it suffices to subtract the appropriate constant from all longitudes. Moreover, the text of the *Almagest* contains a direct confirmation of this explanation: Ptolemy continues: "The latitudinal distances will remain always unchanged, and the positions in longitude (the positions given in the catalog are meant — *Authors*) can provide a ready means for determining the [corresponding] longitude at other points in time, if we [calculate] the distance in degrees between the epoch and the time in question on the basis of a motion of $1°$ in 100 years, [and] subtract it from the epoch position for earlier times, but add it to the epoch position for later times" (Ref. 17, p. 340). *Thus, Ptolemy explains quite clearly how to reduce the catalog to a date by subtracting a constant (thus making the catalog "older" or by adding a constant (thus "rejuvenating" the catalog).* Again, this resembles more a text from a textbook that explains how to reduce star catalogs to particular dates. This book could as well be used in the 16th century, more so because absolute values of longitudes are not needed for constructing star globes; longitudes in this case may be counted from an arbitrary meridian (Ptolemy suggests to use Sirius to this end). Apparently, the absolute values of ecliptic longitudes were never used in ancient and medieval astronomy (and it is difficult to see what they could be needed for). Therefore the reference point for longitudes was chosen arbitrarily. For example, Copernicus in his book *Revolutionibus Orbium Saelestium* (vol. 6) counted longitudes from γ Aries, *distant from the spring equinoctial point by* 27° *(at the time of Copernicus)*.

5. Where did Ptolemy count longitudes from?

Here we will discuss in details the question: From what point of the ecliptic did Ptolemy count longitudes? The traditional opinion is that this was the spring equinoctial point (as late medieval astronomers chose). It turns out however that the problem is much more difficult, and no unique solution follows from the text of the *Almagest*. Let us invoke the corresponding description in the *Almagest*.

Here is Ptolemy's description of the second column of the catalog, that contained longitudes: " ... in the second section, in position in longitude, as derived from observation, for the beginning of reign of Antoninus ([the position is given] within a sign of the zodiac, the beginning of each quadrant of the zodiac being, as before, established at [one of] the solstitial or equinoctial points); ... " (Ref. 17, pp. 339–340). Thus, the longitudes are given separately in each sign of the zodiac, and are counted from the beginning of the corresponding sign. In other words, the catalog contains not the absolute values of longitudes counted from a once and for all chosen point of the ecliptic, but in each case, relative longitudes.

The compiler indicates that the beginnings of one of the quadrants of the zodiac are established at the equinoctial points. Note that the signs of the zodiac do not coincide with zodiacal constellations. Hence, in order to calculate absolute longitudes within a zodiacal sign, we should add integer times 30° (the size of each sign of the zodiac, which are twelve). Only after application of this procedure do we obtain the absolute ecliptic longitudes. For example, in the *Almagest*, the Polar star has longitude denoted as ♊ 0°10'. In order to find the absolute longitude, we must (according to the tradition) add to 0°10' the integer number of grades, namely, 60°; it is accepted that this addend corresponds to the sign Gemini. The sum, 60°10', is supposed to represent the position of the Polar star about the 1st century AD. The procedure is similar for the rest of the thousand stars of the *Almagest*. Despite the simplicity of this calculation, note that it implies the possibility of various interpretations of the original data. The number of degrees to be added to each longitude depends on the choice of the *first* sign of the zodiac, whose rim contains the equinoctial point; Ptolemy's description leaves place for different choices.

Moreover, as we can see further, *Ptolemy is not at all interested in the spring equinoctial point as the zero point for longitudes.* He writes, "Since it is not reasonable to mark the solstitial and equinoctial points on the actual zodiac of the globe (for the stars depicted [on the globe] do not retain a constant distance with respect to these points), we need to take some fixed starting-point in the delineated fixed stars. So we mark the brightest of them, namely the star in the mouth of Canis Major [Sirius], on the circle drawn at right angles to the ecliptic at the division forming the beginning of the graduation, at the distance in latitude from the ecliptic towards its south pole recorded [in the star catalog]" (Ref. 17, p. 405). Thus, *Ptolemy indicates Sirius as the absolute zero point for longitudes.* This is in sharp contradiction with the traditional version that Ptolemy wished to place the starting point at the spring equinoctial point.

Further, more questions arise. The above quotation implies that about a thousand arithmetic operations are to be made by a manufacturer of a globe, namely, subtraction of the longitude of Sirius from the longitudes of the rest of the stars of the catalog. But the longitude of Sirius in the catalog is a fractional number ($17\frac{2}{3}$ grades in Gemini), so the subtraction is a laborious operation. On the other hand, Ptolemy (who suggested to choose the brightest point as the zero point) could also choose the other brightest star, Arcturus. This is a very bright star, and its longitude in the catalog is *integer* (namely, 27° in Virgo). Why subtract a thousand times a fraction if it is possible to subtract an integer? This may mean that the principles of compilation of the catalog differed from the ones that Ptolemy indicates in his description.

The conjecture naturally arises that a constant had been added to the original longitudes (which were meant by the compiler and which we do not know). Moreover, we even can say something about the constant: it makes some integer number of grades and 40'. Why? Because the longitude of Sirius in the available version of the catalog is 17°40'.

Here we come to an unexpected agreement with the results of R. Newton[1]. Based on quite different (statistical) considerations, he proved that the longitudes of the *Almagest* had been recalculated, and in this recalculation, a constant equal to an integer grade 40' was added. This coincidence of results of two different arguments does not seem accidental.

We should make here a general remark, that formally has nothing to do with astronomy, but is probably useful for a fuller understanding of the role and the position of the *Almagest*. A view dominates in modern literature that the chapters of the *Almagest* devoted to particular stars are a kind of comment to the main document, the catalog. However, our impression is different. In fact, the main point here is the description of making a globe, where the stars are to be depicted. The process of making a globe is described in details, for example, what colors are to be used, etc.; as for the catalog, it seems to be just a comment to this description. We cannot exclude that the globes were used, in particular for some mystical and astrological purposes. It is interesting that constructing celestial globes is well-known in the history of astronomy, not only in the 2nd century, but in the Middle Ages, in particular, in the times of Tycho Brahe, who constructed a globe himself. Constructing a globe was considered as an important act. "The big globe is worth attention, 149 centimeters in diameter, with the surface covered by thin brass sheets. The zodiacal belt, the equator and positions of 1000 stars were depicted on the globe, the coordinates of which were determined for the years of Tycho's observations. He remarked proudly that 'No globe of this size and so beautifully manufactured has ever been done anywhere and by anybody in the world ... '. He said also that many specially came to the Dutchman to see this globe. This real wonder of science and art unfortunately was burnt down in a conflagration in the second half of the 18th century" (Ref. 18, p. 217).

6. General review. The blossom of antique astronomy

According to the traditional version of chronology of the history of astronomy, many important astronomic discoveries were made in antiquity. We will only mention briefly some of them. It is considered that in ancient Greece, a handbook on *navigational astronomy* existed, compiled in the beginning of the 6th century BC, presumably by *Thales of Miletus* (ca. 624–547 BC); see Ref. 18, p. 13. As early as in the 4th century BC, *Theophrastus of Athens* observed sunspots (Ref. 18, p. 13). *Meton* (born ca. 460 BC) discovered that 19 years is almost exactly equal to 235 lunar months (the error is indeed less than 24 hours). Almost a century later, *Kallippos* introduced a small correction[2].

Little is positively known about the life of *Pythagoras*. He was born in the beginning of the 6th century BC, and died in its end or about. Pythagoras stated that the earth, like the rest of the heavenly bodies has the form of a ball, and that it floats in the universe without a support. As A. Berry notes[2], the belief that the earth is spherical never was abandoned by Greek philosophers after Pythagoras. *Phylolaes* (ca. 470–ca. 399 BC) outlined a detailed picture of the world in the light of Pythagoras' ideas. He considered that *in the center of the world not is the earth, but the central fire*, and that the earth, the moon and the sun revolve around it, and that the earth also rotates about its axis so that at any moment of time the central fire is invisible for us (Ref. 18, p. 23).

"Phylolaes stated that the distances of heavenly bodies from the central fire raise in geometric progression, so that each successive body is three times further than the preceding. Would he say 'two times', he would anticipate by two thousand years the rule of Tizius and Bode ... " (Ref. 18, p. 31).

As early as in the 6th century BC, *Hicetius*, a disciple of Pythagoras stated the idea that the earth, disposed in the center of the world rotates about its axis with the period 24 hours. *Heraclides of Pontos* (ca. 390–ca. 310 BC) claimed that *Mercury and Venus revolve around the sun*, and around the earth together with the sun (Ref. 18, p. 24). The latest writers indicate three more Pythagoreans, who believed in the motion of the earth, namely, to Hicetius, Heraclitus and Ekthantes, who lived in the 6th and the 5th centuries BC (see Ref. 5). *Democritus* (ca. 460–ca. 370 BC) stated that the universe consists of infinitely many worlds, which arise due to collisions of atoms. These worlds have various sizes; in some worlds there is no sun neither moon, in the other they are greater in size than in the ours, and in the third there are several. Some worlds are completely deprived of moisture, and have neither plants nor animals. Some of the worlds are only dawning, the other are in their blossom, and the third now degrade. "Democritus stated several genial guesses, that found confirmations only several centuries later. In particular, he stated that the sun is much greater in size than the earth, that the moon shines a reflected light, and that the Milky Way is a gathering of a great number of stars" (Ref. 18, p. 25). *Plato* (ca. 428–377 BC) devoted no works specifically to astronomy, but many of his dialogs contain remarks on astronomic subjects. In particular, he considered that the center of the universe is occupied not by the earth, but by a more perfect body[2]. In particular, Plato describes the order of heavenly bodies according to their distances from the earth, which in his opinion looks thus: the moon, the sun, Mercury, Venus, Mars, Jupiter, Saturn. Plato's disciple *Eudoxus* (ca. 408–ca. 355 BC) placed the earth (of course, spherical) fixed in the center of the universe and supposed that the motion of each planet is determined by several spheres (Ref. 18, p. 27); he constructed an extensive theory of these spheres. In particular, Eudoxus explained the inverse motions of the planets and their deviations from the ecliptic. With the help of 27 spheres, Eudoxus explained all visible motions of the planets. *Aristotle* (384–322 AD) stated that *the planets are more distant from the earth than the sun and the moon, and the distance to the sphere of fixed stars is at least nine times as much as the distance from the earth to the sun* (Ref. 18, p. 30). "Aristotle considered most seriously and comprehensively the question about the shape of the earth and the sun. On the basis of the above arguments (lunar phases, the shape of the earth's shade, etc.), he proved that both the earth and the moon are spherical in shape" (Ref. 18, p. 30). Aristotle was aware of the other scientists' arguments that the sun does not move around the earth, but *the earth, together with the rest of the planets revolve around the sun*, but he put forward an objection to this idea: if the earth moved in space, this would bring about regular variation of angular distances between stars, which had been never observed by the astronomers he knew (Ref. 18, p. 30). This argument is quite correct, because *it deals with the really existing effect of the parallactic motion of stars*; the antique astronomers simply could not observe it because of the smallness of the displacements. "Only 2150 years after Aristotle was the yearly parallactic motion of stars discovered ... " (Ref. 18, p. 30).

Among the astronomers of the so-called Alexandrite school, *Aristarchus of Samos*, *Aristyllos* and *Timocharis* are usually mentioned, almost contemporaries, who lived in the first half of the 3rd century BC[2].

The antiquity had "its own Copernicus" (see Ref. 46), *Aristarchus of Samos* (ca. 310–ca. 250 BC), who understood that *several measurements and calculations allow to determination of the distances in the triangle sun-moon-earth*, which he did in his work *On Magnitudes and Distances of Sun and Moon*. His basic assumptions are: 1) The moon takes its light from the sun. 2) The earth is in the center of the motion of the moon. 3) When the moon is seen as bisected, then the large circle dividing the shining and the shaded parts lies in the plain passing through the observer's eye. 4) When the moon is seen as bisected, its distance from the sun is less than one fourth of the circle without one thirtieth of this quarter. 5) The earth's shade is twice as wide as the moon. 6) The moon spans one fifteenth of the zodiac. "This was the first work in the history of astronomy in which the distances between heavenly bodies were determined on an observational basis. Though, the result was very inexact" (Ref. 18, p. 33). Nevertheless, "apparently, these calculations led him to the conclusion that the sun, as a bigger body, is in the center of the world, and that the earth and the other planets revolve about it" (Ref. 18, p. 33). *Archimedes* (ca. 287–212 BC) wrote about this: "Aristarchus of Samos ... comes to the conclusion that the world is much more in size than what is now indicated. He supposes that the fixed stars and the sun do not change their position in space, that the earth moves along a circle about the sun in its center, and that the center of the sphere of fixed stars coincides with the center of the sun, and the size of this sphere is such that the circle supposedly covered by the earth is in the same relation to the distance to the fixed stars as the center of a ball to its surface ... " (Ref. 18, p. 34). *This is practically the point of view of Copernicus.* Aristarchus is also believed to have known the true angular diameter of the moon.

Aristotle measured the earth (as a ball). *Eratosthenes* (ca. 276–ca. 194 BC) obtained a more precise value of the radius of the earth. *It is traditionally believed that the error of Eratosthenes' measurement made 1.3%.* It is believed also that Eratosthenes found the angle of inclination of the ecliptic to the equator, which he estimated as 23°51′. It is interesting that this value was later accepted by Ptolemy in the *Almagest*. As we have noted, this value of inclination of the ecliptic enables us to refine the dating interval for the compilation of the *Almagest*.

S. V. Žitomirskiĭ [40] carried out a reconstruction of Archimedes' model of the world on the basis of his numeric data. As notes I. A. Klimišin, "A well-balanced geo-heliocentric theory appears before the reader's eye, in which Mercury, Venus and Mars revolve about the sun, which together with them, as well as with Jupiter and Saturn move around the earth. The relative radii of orbits of Mercury, Venus and Mars in this model are fairly close to their true values!" (Ref. 18, p. 38). Archimedes created an "automotive instrument", a mechanical celestial globe, with the help of which he demonstrated the conditions of visibility of luminaries, and lunar and solar eclipses. Cicero noted that "the solid sphere without hollows had been invented long ago, and

Thales of Miletus was the first who made such a sphere, and then Eudoxus of Knid, as is said, a disciple of Plato, outlined on it the positions of constellations and stars disposed in the sky, ... many years later Aratus ... praised in verse the arrangement of the sphere and the disposition of the luminaries in it, taken by him from Eudoxus ... The invention of Archimedes is astounding in that he guessed how, having dissimilar motions, to save in one revolution various and different paths. When Gallus brought the sphere into motion, it happened so that on this sphere of brass the moon gave way to the sun in so many revolutions as in how many days it gave way to it in the sky, owing to which the same solar eclipse happened in the sphere, and the moon entered that same balk where was the earth's shade ... " (Ref. 47, p. 14). A similar celestial globe was ascribed to *Posidonius* (after Archimedes). Cicero writes, "If somebody else would bring to Scythia or Britain the sphere made by our friend Posidonius, the sphere whose separate revolutions reproduce what happens in the sky with the sun, the moon, and the five planets in various days and nights, then who in that barbaric countries would doubt that this sphere is a work of a perfect mind?" (Ref. 48, p. 129).

We cannot but recall here the Middle Ages when, say, Tycho Brahe also created his famous celestial globe, apprehended by contemporaries as a real wonder of science and art.

It is considered nowadays that one of the main contributions of Greek astronomy was working out a mathematical point of view on celestial phenomena: spheres of revolution were introduced, the related elements of spherical geometry and trigonometry were worked out, etc. There is a series of small treatises, or handbooks, written mostly in the Alexandrite period and telling about the aforementioned sphere of science (known as spherica, that is, a science about spheres); a beautiful example of such treatise is the *Phenomena* of the famous geometer *Euclid* (about 300 BC). The honor of the author of the guess that the motions of heavenly bodies may be with greater simplicity represented by a combination of uniform circular motions than by Eudoxus's rotating spheres is usually attributed to *Apollonius of Perge* (the first half of the 3rd century BC).

It is now believed that astronomy began to form as an exact science owing to the works of Hipparchus (ca. 185–125 BC). "Hipparchus was the first who started systematic astronomic observations and their comprehensive mathematical analysis; he worked out a theory of lunar and solar motion, a method for prediction (to an accuracy within one or two hours) of eclipses, and founded a basis for spherical geometry and trigonometry"(Ref. 18, p. 43). Hipparchus introduced the distinction between the sidereal year and the tropical year; *he also discovered the phenomenon of precession*, the motion of the spring equinoctial point along the ecliptic towards the sun. 169 years before Hipparchus, astronomers *Aristyllos* and *Timocharis* fixed the positions of eighteen stars in the sky, and Hipparchus used these data to calculate the rate of precession (Ref. 18, pp. 43–44). *Hipparchus compiled a star catalog contain-*

ing 850 heavenly bodies. For each star, he indicated its ecliptic coordinates and the star magnitude. It is now believed that the constellations which he mentions are almost identical with the constellations of Eudoxus (cf. Ref. 2). Nowadays we know about the works of Hipparchus mainly from the Ptolemy's *Almagest*. The only extant work of Hipparchus is his comments to the poem of Aratus and to its source (a work of Eudoxus).

Finally, Ptolemy's *Almagest* is considered to be the final chord of antique astronomy. It is supposed that he was born in Egypt in 127–141 AD, carried out observations in Alexandria, and died about 168 AD.

It is believed today that for three centuries after the death of Hipparchus, the history of astronomy is as if covered by the gloom of centuries of stagnation, in which mainly the discoveries of Hipparchus spread and were popularized. The work of Ptolemy is believed to be a practically unique notable event in the "darkening" history of Greek astronomy.

According to the traditional point of view, Ptolemy was the last prominent astronomer of ancient Greece; after him a period of silence began.

As we see, antique astronomy made many advances, later "rediscovered" by medieval astronomers of the Renaissance. Moreover, the astronomic knowledge was at such a high level in the antique society that it reveals on various occasions, some things of greater nonscientific nature. Thus, some consuls of the regular Roman army could read to their soldiers a real scientific lecture on the theory of lunar eclipses. In decade V of Titus Livius' *Roman History*, a wonderfully precise description of a lunar eclipse is contained: "Consul Sulpicius Gallus declared that 'the next night—let nobody think it is a wonder!—between the second and the fourth hour a lunar eclipse will happen. Since the phenomenon is in the natural order and in a definite time, we can know about it in advance and anticipate it. And so, like we are not surprised that the moon now appears as a full disk, now, on the wane, has the form of a small horn ... so we must not regard as an omen that the moons light is eclipsed when the earth, shade covers it.' In the night on the eve of September nonas, when in the hour specified the lunar eclipse occurred ... " (Ref. 49, X, IV, p. 37).

This detailed lecture (by the way, we only have adduced a part of it) was read before the iron legions of Rome approximately two thousand years ago (see Ref. 50, pp. 190–191, no. 27). The lecture produces a strong impression on a man familiar with the history of science, especially in the light of many peculiarities connected, in particular, with the traditional version of the history of astronomy. Not only consuls, but also Roman architects were great specialists in astronomy. Thus, the level of astronomic knowledge of the *antique architect* Vitruvius surpasses greatly the knowledge of *medieval professional astronomers* up to the 13th century; see, for example, Ref. 4 for details.

Now as we have made certain that the level of the antique astronomy practically corresponds to that of the Renaissance of the 10th–14th centuries, let us turn to the history of astronomy in the next period of time, namely from the 2nd century AD to the 10th century.

7. The thousand years decline in the development of astronomy

After the Roman consul's address to his troops, it is instructive to move to the 6th century AD and listen to the explanation of the structure of the universe delivered by the famous specialist in medieval cosmography Cosma Indicopleustus, who studied particularly the sun and the stars.

He considers that the universe is a box, inside which, on a plane earth washed by the ocean, a great mountain rises. The firmament is supported by four sheer walls of the box—the universe. The sun and the moon go behind the mountain for some parts of the day. The top of the box is strewn with small nails, stars. Four angels stand in the corners of the box, who produce winds. This highly professional point of view is reflected in a medieval printing, reproduced in Figure A3.

The work of Cosma Indicopleustus *Christian Topography* that includes the above point of view is supposed to have been created about 535 AD, and it was *spread very wide in the Christian world*. In order to explain this phenomenon, modern commentators suggest the following version: "If we look at it (the work of Cosma) more attentively, then we see that the *Christian Topography* was spread so widely not because of its ideas about the structure of the universe, but because of the vivid interest of the medieval reader to the beautiful miniatures that decorate the most ancient copies of this document" (Ref. 18, p. 77). As we will see below, this explanation is hardly admissible.

So, what happened? Why this caveman's level of understanding astronomy? Could it be the level of Cosma Indicopleustus only (although he was a recognized authority)? No; in fact, this is an illustration of a general and typical situation.

"*Decline of the antique culture.* After the exciting blossoming forth of antique culture, on the European continent a long period of stagnation, sometimes even of regress, began, usually referred to as Middle Ages ... Over more than 1000 years not a single essential discovery in astronomy was made ... " (Ref. 18, p. 73). The traditional explanation is that Christianity was incompatible with science.

In A. Berry's opinion[2], the history of Greek astronomy ends with Ptolemy: in eight and half centuries that separate Ptolemy from Albatenius almost no observations of any scientific value were done.

"*A relapse of childhood.* The idea about the plane earth originated in the time of childhood of humankind ... But we have already seen that Greek philosophers could produce scientific proofs that the earth is spherical, determine its size and, though not very precisely, the distances to the sun and

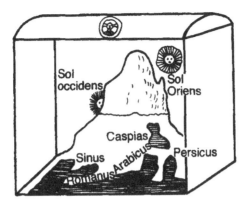

Figure A3.   The world according to Cosma Indicopleustus (the 6th century AD).

the moon ... But the new generations, overcome by religious fanaticism ... destroyed the building that began to be erected. Here and there ... relapses of childhood occur in the views of the world. In particular, the ideas about the plane earth were revitalized for many years (up to the 9th century!)" (Ref. 18, pp. 74–75).

As A. Berry notes[2], about fourteen centuries passed from the publication of the *Almagest* to the death of Copernicus (1543). In this period not a single astronomic discovery of paramount importance was done. Theoretic astronomy hardly advanced at all, and in some aspects even regressed, since the prevailing doctrines, in some respects more correct than Ptolemy's, were professed with much lesser understanding and conscience than in antiquity. As we have already seen, in the west, nothing remarkable happened in the first five centuries after Ptolemy. Then an almost overall blank space begins, and still many centuries passed before a revival of interest to astronomy.

The conclusion of Berry is that in Europe the troubled period that followed the fall of the Roman empire in the 6th century AD and preceded the establishment of feudalism appears to be a blank spot in the history of astronomy, as well as generally in the history of science. The best minds of the epoch were buried in theology, if not in practical activities.

8. The repeated blossom of astronomy in the Renaissance

1. Arabic Renaissance of Astronomy

This scientific movement in Islamic countries can be but with major reservations called the true renaissance of the ideas of antiquity. As A. Berry notes[2], no one Arabic astronomers (we will list them below) gave rise to a significant original idea. Instead, they had a perfect ability to assimilate somebody else's ideas and give them a further, though minor development. They were all patient and thorough observers and skilled calculators. We owe them for a long rank of observation and inventions, as well as for improvements of mathematical techniques. The Arabic "renaissance" looks more like a *dawning* of astronomy as a science. The "long rank of observations", which initiates any exact science agrees with this quite well. For reader's convenience, we adduce here some chronological data about the most important activists of the Arabic renaissance of astronomy.

It is traditionally believed that the first translation of the *Almagest* was done after the order of the successor of Almanzor, Harun Al Rashid (765 or 766–809), the hero of *Thousand and One Nights*. Apparently, this was not easy; the other attempt to translate the *Almagest* was done by Hunayn ibn Ishāq (?–873) and his son Ishāq ibn Hunayn (?–910 or 911), and the final version, established by Ṭābit ibn Qurra (836–901) appeared by the end of the 9th century. To this activity of Arabs we owe the preservation of many Greek works, whose original manuscripts did not survive. By the way, the original of the *Almagest* is also believed to be lost.

In the period of caliphs' residence in Damascus, an observatory was built there. Another observatory was built in Baghdad by caliph Al Ma'mūn in 829. Al Ma'mūn charged his astronomers to verify Ptolemy's estimate of the size of the globe. Two independent measurements of a part of meridian were done, which, however, agree so well with each other and with Ptolemy's erroneous result, that they can hardly be treated as independent and thorough measurements; most probably, these should be regarded as a rough check of Ptolemy's calculation. On the other hand, so much attention was given to the accuracy of measurement that according to some data, the most important of them were fixed in formal documents, sealed by a joint oath of several astronomers and lawyers (see Ref. 2).

In the second half of the 9th century, in Baghdad worked *Akhmed Al Fargani (Alfraganus)*, the author of *Elements of Astrology*, and *Ṭābit ibn Qorra* (836–901) (see Ref. 2, p. 80). It is interesting that in this time astronomic tables start to appear, constructed on the same principle as the *Almagest*. To Ṭābit ibn Qorra belongs the doubtful honor of discovery of a purported variation of precession. Wishing to explain it, he invented a complicated mechanism and introduced an arbitrary complexity that tangled and obscured a majority of astronomic tables that appeared the next five or six centuries[2]. *Al Battānī (Albatenius)* is believed to have been a much more skilled astronomer, who made observations in 878–918 and died in 929. The last Baghdadian astronomer was *Abū'l-Vafā'* (930 or 940–998), the author of a voluminous treatise in astronomy, radiant with new ideas and written in a different manner from Ptolemy's book, although, as A. Berry states, equally popular; it was often mistaken for the translation of the *Almagest* (!). Can it be that the *Almagest* originated in the works of Abū'l-Vafā'? *Ibn Yunus* (?–1008 or

950–1008) is almost a contemporary of Abū'l-Vafā' (Ref. 18, p. 83). He created astronomic and mathematical tables (the so called *Hakemit tables*) that served as a pattern for two centuries. An outstanding phenomenon of medieval observational astronomy is the *Book of Fixed Stars* of *As-Sûfi* (Abd ar Rahman As-Sûfi, 903–986). The name *As-Sûfi* means simply "wizard" (Ref. 18, p. 80). The book contained many illustrations and a star catalog. It is believed that As-Sûfi "verified and refined the Ptolemy's star catalog" (Ref. 18, p. 80).

Abu al-Raihan Bīrūnī (973–1048) made independent astronomic observations, calculated the angle of inclination of the ecliptic to the equator, obtaining the value 23°33'45". He constructed "apparently, the first in the world" (Ref. 18, p. 83) terrestrial globe (more exactly, a half globe) five meters in diameter. In 1031–1037, Bīrūnī created his *Canon of Mas'ud*, an encyclopedia of astronomy. In this book he gives a different value for the inclination, 23°34'0" (the true value at that time was 23°34'45"); the book contains a star catalog embracing 1029 stars, for which the coordinates and magnitudes are given according to Ptolemy and As-Sûfi (Ref. 18, p. 84). "In the whole, *Canon of Mas'ud* follows the plan of the *Almagest* and is in the spirit of geocentrism" (Ref. 18, p. 84).

In the 10th–12th centuries the astronomers that worked in the islamic part of Spain made considerable advances. The astronomer *Al-Zarqālī*, known also as *Arzakhel* (1029–1098), improved the astrolabe and published in 1080 a volume of astronomic tables (the so called *Toledan tables*). Also, various problems of astronomy were considered by *Muhammed Ibn Rushd*, or *Averrhoes* (1126–1198), *Mozes Ben Maimon*, or *Maimonides* (1135–1204), then *Al Bitruji* (died about 1204). It is considered that Al Bitruji "revived" some ideas of Eudoxus (Ref. 20, p. 86). As A. Berry writes, some improvements of instruments and techniques of observation are due to this school; it published several treatises devoted to a critique of Ptolemy, that, however, introduced no corrections to his ideas. Meanwhile, the Christian Spaniards began little by little to dislodge their Islamic neighbors. Cordoba was seized in 1236, Seville in 1248, and after their fall the Arabic astronomy perished for history[2].

The next center of development of astronomy is usually connected with the reign of Gulagu Khan, the grandson of Genghis Khan. In 1258, he seized Baghdad. Several years before, he favored an astronomer *Nassir Al Din Al Tusi* (*Nassir Eddin*, 1201–1274). He established a significant scientific center and an observatory in the town Maragha (now in Iranian Azerbaijan). The instruments they used were characterized by their size and solidness of construction and were, apparently, better than that used in Europe in the times of Copernicus; only the instruments of Tycho Brahe first surpassed them. The work of the astronomers yielded a series of astronomic tables based on the Hakemit tables of Ibn Yunus and called the *Ilkhan tables*. The collection contained tables for calculation of planetary positions and a star catalog, to some extent created from new observations.

An outstanding astronomic center in the 40 year long reign of *Ulugh Beg* (1393–1449), grandson of Timur, was in Samarkand. Here in 1424, a large observatory was built. Ulugh Beg published new planetary tables, but his main work was the star catalog, containing almost the same stars as Ptolemy, but using new observations. This was most probably the first completely independent catalog after Hipparchus. The positions of stars are given with an extraordinary accuracy, not only in grades, but also in minutes, and though in comparison with modern observations errors about several minutes are usual, the instruments used by Ulugh Beg were apparently very good. Together with him, Tartar astronomy faded[2].

If we prescind for a moment the idea that a blossom of antique astronomy preceded these investigations of Arabic astronomers, we will have to admit that the Arabs set forth new deep ideas. In this case, the opinion of A. Berry which we adduced in the beginning of this subsection will turn out to rely on the traditional chronology, which places the achievements of the antique astronomy long before the "Arabic astronomic renaissance".

2. European renaissance of astronomy.

In the 10th century, the fame of the Arabic science spread little by little via Spain to other parts of Europe. The famous scientist *Herbert* (who was the Pope under the name *Sylvester II* from 999 until 1003) was especially interested in mathematics and astronomy. "Many other scientists were interested in Arabic science, but only a century later did the influence of Mohammedans become perceptible. In Byzantium, only as late as in the 9th century did *Michael Psellos* (1018–ca. 1097) and *Simeon Setus* "revive" and begin to bring into order numerous (and ostensibly, in the traditional chronology, known since the time of Aristotle) proofs of the sphericity of the earth, discuss the length of the earth's equator, the ratios of the radii of the sun, the earth, the moon etc. (Ref. 18, p. 78).

"A significant role in the raise from the *millennium long lethargy* belongs to Italy" (Ref. 18, p. 92). It is believed that *in the beginning of the 12th century*, translations of scientific and philosophic works from Arabic into Latin began to appear. *Plato of Tivoli* translated the *Astronomy* of Albatenius about 1116; *Adelard of Bath* translated Euclid's *Elements*, *Gerard of Cremona* (1114–1187) translated, in particular, the *Almagest* and the *Toledan tables* of Arzakhel[2]. An interest in the works of Aristotle arose among European scientists in the 11th–12th centuries, and in the 13th century their influence on the medieval science becomes almost dominating; many scholastics respected his works equally, if not more than the creations of the most prominent Christian theologians[2]. The acquaintance of Western Europe with the Arabic astronomers developed in the reign of Alfons X (king of Leon and Castilla, 1223–1284), under whose supervision a group of scientists calculated a series of new astronomic tables, the so called *Alphonsine tables*, which took the place of the *Toledan tables*.

The *Alphonsine tables* were published in 1252 and spread fast in Europe. It is considered now that they did not contain any new ideas, but many numeric data, in particular the length of the year, were determined with greater accuracy than before[2]. In the reign of Alfons, the book *Libros de Saber* was created, a voluminous encyclopedia of astronomic knowledge of the time, which, although taken to a great extent from Arabic sources, *was not at all, as was sometimes believed, a mere inverse translation.* In this book, a picture showed the orbit of Mercury as an *ellipse* with the earth in the center; this was probably the first idea that the orbits of heavenly bodies could differ from circumferences.) The *Alphonsine tables* were "in use in all European countries for two centuries" (Ref. 18, p. 93).

The English astronomer *John Halifax* (ca. 1200–1256) is more famous under the latinized name *Sacrobosco*. His treatise *Sphaera Mundi* was very popular for three or four centuries; it was often republished, translated and commented upon; this was one of the first printed books in astronomy; 25 editions of this book appeared between 1472 and the end of the 15th century, and 40 more editions appeared about the middle of the 17th century[2].

Nevertheless, if we assume that the advances of antique and Arabic astronomy date before the 10th–12th centuries, then we must come to the conclusion that the scientists of the 10th–13th centuries were content with collecting and bringing into order the parts of Greek and Arabic astronomic science that they could master; we do not see here any serious attempts to develop a theory, neither any important observations.

The famous French scientist *Jean Buridan* (ca. 1300–ca. 1358) is the author of a book about the structure of the universe; in particular, he studied in details the question whether the earth is always fixed in the center of the universe. His disciple *Nicholas of Horem* (ca. 1323–1382) published *Book of the Sky and the Universe*, in which he supported (as a hypothesis) the idea of the diurnal rotation of the earth. *Nicholas of Kues* (1401–1464) stated that the earth cannot be in the center of the universe (Ref. 18, pp. 96–97).

It is believed that only as late as in the 15th century a German school appeared, that contributed to astronomy[2]. *Georgius Peurbach* (1423–1461) wrote *Concise Exposition of Astronomy*, presumably based on the *Almagest*; it is believed, however, that he used *bad* Latin translations, "full of errors" (Ref. 2, p. 84). The activities of Peurbach were continued by his disciple *Iohann Müller* (or *Wolfghang Müller*; Ref. 18, p. 94), also called *Regiomontanus* (1436–1476). Both astronomers made many observations. It is believed nowadays that "Peurbach was the first in Western Europe who expounded Ptolemy's theory in combination with Aristotle's system of the world" (Ref. 18, p. 94). However, this book of Peurbach (*New Theory of Planets*) was only published by Regiomontanus in 1472, after the death of Peurbach. Then Regiomontanus supplemented Peurbach's *Concise Exposition of Astronomy*; he published it in 1472 or 1473, using his own printing machine (in Nürnberg). It is traditionally believed that after the death of Peurbach in 1461, Regiomontanus went to Italy, where he got an opportunity to read the *Almagest* in Greek, and in 1468 returned to Vienna (with some Greek manuscripts), and then went to Nürnberg, where he was accepted with great honors. A rich man *Bernard Walter* (1430–1504) supplied him with considerable sums of money and became a disciple and collaborator of Regiomontanus (although he was much older). The most skilled craftsmen of Nürnberg made astronomic instruments with precision never before known in Europe, although the instruments were most probably not as good as the instruments of Nassir Eddin and Ulugh Beg (not extant, but traditionally believed to have been created several centuries earlier). After the death of Regiomontanus in 1476, Walter continued the works of his friend and made a series of good observations; he was the first who successfully introduced errors for atmospheric refraction, of which Ptolemy, apparently, had but a vague idea. "Using Ptolemy's description of the instrument, he made an armilla, with the help of which he determined the positions of planets to an accuracy 5′, and the altitude of the sun to 1′, which is much more accurate than Ptolemy's observations" (Ref. 18, p. 95). At the time, as is now believed, the astronomic instruments known since the time of Ptolemy came into broad scientific use. *Leonardo da Vinci* (1452–1519) was the first who gave the correct explanation for the faint shining observed in the dark part of lunar surface when its lighted part is in the phase of crescent [2]. *Jerome Fracastor* (1483–1543) and *Petrus Apian* (1495–1552) first noted that the tails of comets are always directed to the side opposite to the sun; they are also the authors of several famous astronomic books. *Peter Nonius* (1492–1577) found correct solutions for several problems about the length of twilights. A new measurement of the size of the earth, the first since the time of caliph Al-Ma'mûn, was made about 1528 by French doctor *Jean Fernell* (1497–1558).

We have come to the time of Copernicus. Klimišin notes in his book, "Thus after the millennium long period break, astronomic observations and quests for the laws of the universe started again. In the whole, as was noted by English mathematician and astronomer Edmund Whittaker (1873–1956), 'in 1500 Europe knew less than did Archimedes, who died in 212 BC' ... " (Ref. 18, p. 98).

It seems relevant to tell here about the history of the acquaintance of European scientists with the works of Euclid, Archimedes and Apollonius, because, as we see from the above survey, just in the Middle Ages all the ancient scientific advances "revived".

M. Ya. Vygodskiĭ writes: "Not a single antique manuscripts of Euclid's *Elements* came down to us ... The most ancient available manuscript is a copy made in 888 AD ... There is a lot of manuscripts related to the 10th–13th centuries" (Ref. 51, p. 224). I. G. Bashmakova communicates that still before publication of the first Latin translation of Diophantus' *Arithmetic*, European scientists "used algebraic methods of Diophantus, not being familiar with his works" (Ref. 52, p. 25). She characterizes this situation as "somewhat paradoxical". The first edition of *Arithmetic* dates back to 1575 AD.

Copernicus immediately followed the *Almagest* (recall that the rise of interest in the *Almagest* and its publications occurred in the time immediately before Copernicus), and Diophantus was equally immediately followed by Fermat (1601–1665). The history of manuscripts and printed editions of Archimedes follows the pattern we have already seen. I. N. Veselovskiĭ communicates that the *base* for all modern editions of Archimedes is *a lost manuscript of the 15th century* and the so-called *Constantinople palimpsest*, found only *in 1907*. It is believed that the first of Archimedes' manuscripts appeared in Europe *after 1204 AD*. The first translation was made in 1269, but *only found in 1884*. The first printed edition appeared in 1503, the first Greek edition in 1544, "and only then the works of Archimedes come into use in the scientific world" (Ref. 53, pp. 54–56). Apollonius' *Conics* were first published *only in 1537*, and "Kepler, who was the first to discover the significance of conic sections (ellipses) did not live till the first edition of works of Apollonius. The next three books ... were first published in Latin translation (and again translation—*Authors*) in 1631" (Ref. 43, p. 54). Thus, the work of Apollonius was first completely published *after* Kepler's discovery of the importance of the objects discovered therein.

3. The blossom of European astronomy.

Nicholas Copernicus (1473–1543), the author of the heliocentric system of the universe, stands (as believed) at the start of a rapid independent blossoming forth of European astronomy[2]. We have already noted in Chapter 1 the continuity of ideas and "astronomic observations made with an interval almost 2000 years long: as he considering the problem of precession, Copernicus adduces the observational data of his far predecessors ... " (Ref. 18, p. 109). Copernicus referred to Timocharis, Hipparchus, Menelaus, Ptolemy, Albatenius and other. The theory of Copernicus was propagandized by *Reticus* (*Georgius Ioachim*, born in 1514). After him, the famous astronomer who accepted the new views at once, was his friend *Erasmus Reingold* (1511–1553)[2]. Reingold calculated on the basis of Copernican theory and published tables of motion of heavenly bodies (the *Prussian tables*). These tables turned out to be better than the *Alphonsine tables*, and they were popular until the appearance of Kepler's *Rudolfian tables*. In 1561, *Wilhelm IV of Kassel in Hessen* (1532–1592) built the Kassel observatory, where he, together with young skilled astronomers *Christian Rothmann* and *Jost Bürgi* started a compilation of a star catalog (see Chapter 1 and Ref. 2). By 1586, positions of 121 stars had been most accurately measured. Then, the work of *Tycho Brahe* started to dominate in the development of medieval astronomy; see Chapter 1 for details about the works of Tycho Brahe. In the 21 years spent by Tycho on the isle Gwain, he collected with the help of his disciples and assistants, a series of brilliant observations, surpassing in accuracy and extent all that had been done by his predecessors. Much attention he gave to alchemy and, to some extent, medicine.

Further, the development of astronomy became so large-scaled and branched that we have no possibility to consider all its directions in this review (and in fact our purposes do not require that). Therefore, we will confine ourselves to a brief enumeration of the most prominent scientists and their main advances. Starting here, we will redirect our attention to the large chronologic table, to which the next section is devoted.

Giorgiano Bruno (whose real name is *Philipp*; 1548–1600) fought for the idea of boundlessness of the universe and of multiplicity of worlds; he is the author of several books on philosophy, in which he actually developed the ideas of Copernicus.

Galileo Galilei (1564–1642) is a very famous astronomer, the author of many extremely important astronomic discoveries: the start of telescopic observations, the moons of Jupiter, the phases of Venus, and other. Galileo was an active advocate of the Copernican system.

Johann Kepler (1571–1630), a disciple of Tycho Brahe, discovered fundamental laws of planetary motion.

In the 17th century, the first measurement of the earth was made, which was a decisive step in comparison with the measurements made by Greeks and Arabs[2]. The following names are connected with these measurements: *Willebrord Snellius* (1620–1682), *Richard Norwood* (1590?–1675), *Jean Picard* (1620–1682).

Here we stop and pass to the exposition of our next idea that allows us to display visually the described development of astronomy and views of the structure of the universe with time.

9. Chronological table and the diagram of development of astronomy with time

In the interval 10th century BC to 20th century AD, let us mark the centuries and try to depict the qualitative picture of development of astronomy from antiquity to nowadays. As the material to be depicted, we take the dates of activities in astronomy of various scientists; for each of them, we mark in the diagram the horizontal segment whose ends correspond to the dates of his birth and death. The number of the segments that meet an interval of time will show visually the intensity of development of astronomy in the period. Of course, this principle is but provisional, but its advantage is due to the fact that a concrete astronomic information is connected with each name in the history of science, so we can trace its evolution in our diagram. The "number of astronomers" in this or that epoch is a rough indicator, but it shows more or less the intensity of development of the science. It is as well desirable to consider the "quality of the science" produced in a period; to some extent, we have done this in the previous section.

The next question is how to compile the list of astronomers from antiquity to nowadays. Clearly, it is impossible to create a *complete* list, so we took the astronomers mentioned in three monographs.[1,2,18]. R. Newton's book[1], together with a study of the *Almagest*, exposes a remarkable review of antique and, partially, of medieval astronomy, and Refs. 18 and 2 are devoted to the history of astronomy from antiquity to our days. The main attention is given in these books to the following three categories of historical persons:

1) Astronomers (professional astronomers, observers, etc.);

2) Philosophers and writers who discussed astronomic observations, phenomena and theories;

3) Commentators of astronomic works and translators of astronomic books.

We attended these three categories, and have written out all (literally!) names belonging to the three categories that occur in these books. In those cases when the books gave no chronologic data about this or that person, we used encyclopedic editions.

Doing this, we used all of Ref. 1; as for Ref. 2, we only processed pages 17 to 244 (we were not interested in the contemporary period), and in Ref. 18, for the same reason, we used pages 5 to 189. In other words, we gathered here all information related to the period since antiquity until the 18th century inclusive. From the end of the 18th century, the number of astronomers grows rapidly, and we found no reasons for figuring out any statistics here.

Let us now explain why we have chosen this way of compiling the table.

Of course, the three books were not supposed to contain a complete list of all the names of the three categories. But the authors obviously did their best to reflect the history of development of astronomy fully. The selection they have made may therefore be treated as the effect of a "mechanism of ordering information"; as a result, first the most famous names are mentioned, then the less famous, then almost unknown. Of course, some names do not pass at all; we may assume that either these names are related to the details totally unknown to the modern history of astronomy, or their activity is not treated as deserving to be mentioned in reviews of this type. Not trying to understand very deeply the mechanics of this "sifting out and forgetting", we may assume nonetheless that it is of a "uniform" nature; it models the "sifting out" that the history of science elaborates with time (forgetting one names and keeping in memory the other).

Besides, we have deliberately taken *three* books, but did not confine ourselves to one; thus we tried to level out biases of the authors.

A reader wishing to get a fuller acquaintance with the features of the "forgetting mechanism" may turn to Refs. 8 and 13; we do not dwell on these

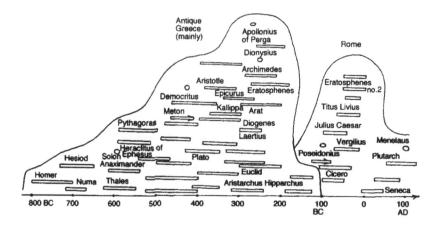

Figure A4. Distribution in the time axis of dates of lives of well-known astronomers, commentators of astronomic works, translators of astronomic books, and philosophers and scientists involved with astronomical problems (according to the traditional chronology).

interesting questions here, because we are not going to use this information in this book.

We did not try to put the names into strict chronological order, but we tried, when possible, to order the names with respect to the dates of birth (which, unfortunately, are not always available). It turned out that the names divide naturally into several disjoint groups, corresponding to different geographic regions. Correspondingly, our list decomposed into groups. Here is the list of the groups; for each group, we give the numbers of the names that we placed in this group.

Ancient Greece:	names 1–37
China:	38, 39
Babylon:	40,
Rome and Europe, the 2nd century BC–700 AD:	41–75
India:	76
Byzantium:	77–82
Islamic countries:	83–108
Europe, 700 AD–the 18th century:	109–220
Total: 220 names.	

In Figures A4, A5 and A6 we depict the dates of the lifetimes of all the 220 persons. Because of the lack of room, we did not write down all the names; all antique names, and the most famous medieval are written.

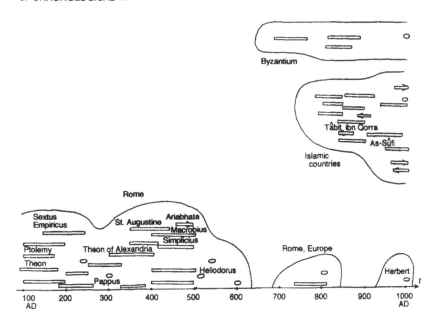

Figure A5. Continuation of Figure A4.

The medieval "period of regress" in the development of Rome and Europe is seen very clearly in the diagram; it affects even the number of the names related to astronomy, not to mention the "quality" of the period (see above). Only in 1100 AD, a gradual progress begins. It is obvious that the "Byzantine" and the "Islamic" parts of the diagram are quite sharply localized in time. The Byzantine "renaissance" starts in the 7th century AD and ends in the 9th century, and the "Arabic peak" begins in the 8th century and ends in the 12th century (the "density of astronomers" decreases rapidly after the 12th century). In order to make this more visual, let us draw the "graph of density", counting for each century the number of astronomers who lived (at least some time) in this century. Of course, the same person may live in two subsequent centuries. The graph is shown in Figures A7 and A8. In Figure A7, the solid line depicts the "density" in the Islamic countries, and the dotted line in Byzantium. The local character of these outbursts of interest in astronomy is obvious. As we have noted, the peak of the "Arabic astronomic renaissance" falls in the 9th–11th centuries. Figure A8 shows the levels of interest in astronomy in Ancient Greece, Rome and Europe. The antique period emerges sharply (the peak in the left side of the graph), then we see the medieval stagnation (especially in the 7th–11th centuries), and since the 12th century, a rapid growth. Note that the graph shows a monotone increase since the 13th century until now; no peaks or oscillations occur in this period.

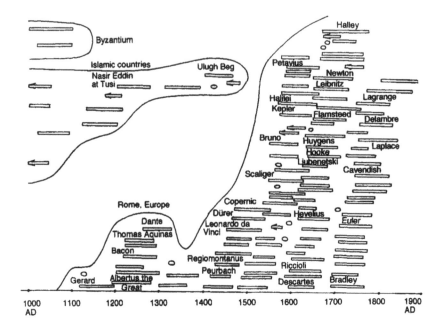

Figure A6. Continuation of Figure A4.

Conclusions.

1) In traditional history of astronomy, a strange phenomenon takes place: a wonderful peak of the antique astronomy, followed by a thousand year long regress, after which another rise starts in the 13th century.

2) Practically all achievements of medieval astronomy in the 11th–14th century had been (according to the traditional chronology) reached more than a thousand years earlier by antique astronomers.

Let us list here some principal ideas discovered by antique astronomers and "rediscovered" only in the 11th–16th centuries AD.

1) Ecliptic and equatorial coordinates and the methods of recalculations from each to the other.
2) Determination of principal elements of motions of planets.
3) Heliocentric theory.
4) Determination of distances in the moon-earth-sun system.
5) Prediction of eclipses.
6) Compilation of star catalogs.
7) Constructing celestial globes.

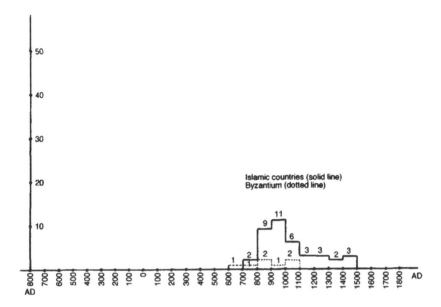

Figure A7. Distribution in time (in the traditional chronology) of the number of astronomers in Islamic countries (solid line) and in Byzantium (dashed line).

8) Discovery of precession.
9) Manufacturing instruments: astrolabes, etc.
10) Computation of the sidereal year and the tropical year.
11) Distinguishing constellations (and fixing them in a picture).
12) Raise of the question on proper motion of stars.

Modern chronologists also try to make us believe that Chinese astronomers determined by measuring the length of the shade of a gnomon, the angle of inclination of the ecliptic (the estimate obtained was 23°54′02″) *about 1100 BC*, long before the astronomic "peak" in Greece[18].

Not stating this far any conjectures, we must note that these facts appear to be quite strange; the strange things in fact only arise from the traditional version of the chronology of antiquity.

10. Some other peculiarities in the *Almagest*

In this section we demonstrate briefly some other peculiarities of the star catalog of the *Almagest*, first indicated by A. N. Morozov in Ref. 4, vol. 4. Not

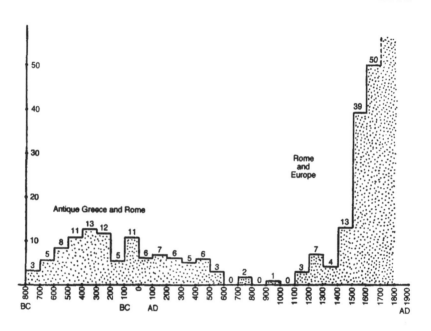

Figure A8. Distribution in time (in the traditional chronology) of the number of astronomers in antique Greece and in Europe.

all statements of Morozov pass indisputably, so we confine ourselves to the ones that we have verified ourselves.

1. In what coordinates was the star catalog of the Almagest originally compiled?

We know that the available versions of the *Almagest* give *ecliptic* coordinates of stars. Above, we have adduced many arguments for the statement that the catalog had originally been compiled in *equatorial* coordinates, and then rearranged (via analytic recalculation or in some other equivalent way) into the catalog of ecliptic coordinates. It turns out that the traces of the rearrangement may be detected.

Since the compiler of the catalog describes the stars of the Northern hemisphere (he starts from northern constellations and descends step by step to the south), it would be natural that he had begun his catalog with the description of the constellation that lies *in the center of the hemisphere, near the pole of the ecliptic*.

The constellation nearest to the pole of the ecliptic is Draco. For the last two thousand years (from the 1st century AD, when the catalog was ostensibly compiled), the pole of the ecliptic did not shift much in comparison with

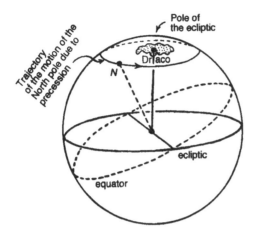

Figure A9.   Trajectory of the motion of the North pole due to precession.

the dimensions of constellations; so, the compiler should be expected to have started the catalog from Draco (in whatever time since antiquity until nowadays he lived). *But the catalog of the Almagest starts not from Draco, but from Ursa Minor.* Then the author describes Ursa Major, and only after it, Draco.

All comes into order if we go back to the equatorial coordinate system. The fact is that in the last three thousand years there really was a period when Ursa Minor was the nearest to the pole constellation. Thus, the compiler of the catalog, as he starts from the stars of Ursa Minor, in fact, gives us the original form of the catalog: the catalog started from the pole of the equatorial coordinate system (Figure A9).

N. A. Morozov writes, "But why in this case did not he leave the immediate equatorial magnitudes as is done in all modern star catalogs, but recalculated in a laborious graphic way to ecliptic latitudes and longitudes?.. Thus he inevitably introduced an additional error and diminished the merit of his catalog ... All the enormous work made by the author in order to recalculate graphically the catalog of fixed stars from the original equatorial coordinates into ecliptic ... was so tremendous and obviously harmful for the astronomic accuracy, that it makes us unwillingly look for an external reason, and there could be two such reasons: either a wish to make the catalog eternal (which failed because of longitudes), or a deliberate urge to mask the time of compilation, because before Newton and Laplace, the ecliptic coordinates were believed to be forever invariant ... "(Ref. 4, vol. 4, p. 201).

Another natural questions arises at once. The northern pole moves notably in the firmament with time. Knowing the laws of this motion, can we use it for a refinement of the dating interval for the catalog?

2. *Polar star as the first star in the* Almagest.

The catalog of the *Almagest* starts from the Polar star. At first sight, this is quite natural. Indeed, if the catalog describes the stars of the Northern hemisphere in equatorial coordinates, then it is natural to start from the star nearest to the North pole, the center of the Northern hemisphere. But even a superficial reasoning on the subject gives rise to puzzled questions.

Modern traditional chronology tries to prove that the catalog of the *Almagest* was created about the 2nd century AD (or a little earlier, in the time of Hipparchus). It is possible to compute that in the last two and a half thousand years, Ursa Minor was (and is) the constellation nearest to the North pole. Further, it is possible to determine, which *star* of Ursa Minor was the nearest to the pole about the 1st century AD. *It turns out that this was β Ursae Minoris.* Furthermore, this star is marked in the catalog as of the second star magnitude, that is, brighter than the Polar star (described in the *Almagest* as the star of the third magnitude). Notice by the way that the *Almagest* does not use this notation (α, β, etc.). The stellar positions are determined by Ptolemy in relation to the rims of constellations (see below for more details). Note that in fact, the stars α and β Ursae Minoris *are practically equally bright*; according to modern photometry, α has magnitude 2.1, and β 2.2, so α is just a bit brighter than β, but Ptolemy thought vice versa (see Ref. 22, p. 51, catalog no. 2). A calculation shows that in the 2nd century AD the distance of β from the North pole was about $8°2'$, while the Polar star (that is, α Ursae Minoris) was $12°1'$ far from the pole. *Thus, the Polar star was much further from the pole than β Ursae Minoris!* The reader can see the position of these stars in Figure A10, which reproduces a part of the star map constructed according to the catalog of the *Almagest* by the famous astronomer Bode (of course, the positions of stars and constellations appropriate for the 2nd century AD are calculated and shown here, because Bode had no doubts about the time when Ptolemy lived; see Ref. 26, p. 238). Further, the star β is in the center of the trunk, but α is on the tip of the tail of the Little Bear, and this is how the positions of these stars are described in the *Almagest*. The Polar star is described in the following words: "The star on the end of the tail", and β as "The southern star in the rear side" (Ref. 17, p. 341). As can be seen from the map in Figure A10, β is at the center of the trunk, nearer to the back, that is, to the top of all the figure (if we turn the Little Bear its paws down). Let us gather the facts about the two stars in Table Ad. 1.

Comparing the columns of the table, we must admit that it is psychologically impossible that the catalog was compiled in the 2nd century AD. A. N. Morozov wrote, "Who in the 2nd or the 3rd century could have an idea to start count in a description of the sky from the star in a northern constellation which is the most distant from the pole, and is not in the middle of the trunk of the Little Bear, where the nearest to the pole star was at that time, but from the tail, where the most distant at the time star was disposed"

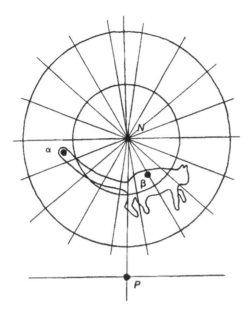

Figure A10.   β Ursae Minoris in the star map of Bode.

Table Ad. 1

The Polar star (α Ursae Minoris)	β Ursae Minoris
1. Described in the *Almagest* as a star of the third magnitude, that is, a fainter star than β (in fact, is equally bright with β).	1. Described in the *Almagest* as a star of the second magnitude (and is considered to be one of two brightest stars in the constellation, because only β and α are described as of the second magnitude, and the rest as being fainter).
2. In the 2nd century AD was far from the pole, at the distance 12°.	2. In the 2nd century AD, was at 8° from the pole, nearer than α.
3. Described in the *Almagest* as "the star on the end of the tail".	3. Described in the *Almagest* as disposed on the top of the Bear's trunk, in the very center of the constellation.

(Ref. 4, vol. 4, p. 202). The situation becomes still more strange if we assume that the *Almagest* was compiled by Hipparchus in the 2nd century BC.

The situation alters sharply, and the strange things disappear if we abandon the hypothesis that the *Almagest* was compiled about the 1st century AD. Let

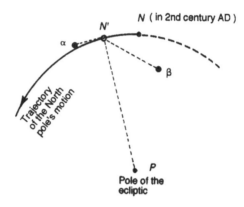

Figure A11. Positions of α and β Ursae Minoris in relation to the trajectory of apparent motion of the North pole.

us see if an epoch exists when starting a catalog from the Polar star was quite natural.

We depict in Figure A11 the North pole N, the pole of the ecliptic P, the stars α and β of the Little Bear, and mark the direction of revolution of the North pole (around the pole of the ecliptic; we neglect oscillations of the ecliptic here). It is obvious that the situation alters with time: β moves away from the North pole, while α, to the contrary, approaches the pole; the North pole moves practically right towards α. The starting position N of the pole shown in Figure A11 is the position in the 2nd century AD. The revolution of the pole around the pole of the ecliptic has the velocity about 1° a century (roughly), so we may estimate the time when the North pole was nearer to the Polar star than to β. We did not try here to make any precise calculations, because we do not treat this dating method as essential, and only regard this argument as auxiliary; a very rough estimate shows that only seven to nine centuries later than the 2nd century AD does α approach the North pole to a less distance than β. Thus, starting from approximately the 9th–11th centuries until nowadays, we have the situation as in Table Ad. 2.

It is quite clear that an observer who compiled his catalog after the 9th century, would choose the Polar star to start the list. This is what the compiler of the *Almagest* did.

For example, in the 15th–16th centuries (when, by the way, the manuscripts of the *Almagest* were published most actively), the Polar star was already less than 4° distant from the North pole, and no other star was nearer. (In 1900, the distance from the Polar star to the North pole was 1°28′, and in 2100 the distance will reach 28′ and will start to increase.)

Table Ad. 2

The Polar star (α Ursae Minoris)	β Ursae Minoris
1. The nearest in Ursa Minor to the North pole.	1. At a greater distance from the pole than α.
2. The tail of the Little Bear is a part of the constellation nearest to the pole.	2. The trunk moves (together with β) away from the North pole.
3. α is brighter than β. It has star magnitude 2.1, and is the brightest star in the constellation.	3. The true magnitude of β is 2.2, so it is fainter than α (although Ptolemy claimed the contrary).

Thus, we have to conclude that the compiler of the catalog, as he started his description from the Polar star, gave out a temporal bound for his observations — not before approximately the 9th century AD.

3. Peculiarities of the Latin (1537) and Greek (1538) editions of the Almagest

The above data about the printed editions of the *Almagest* bring about the conclusion that two editions are most important, the Latin edition of 1537 (Cologne) and the Greek edition of 1538 (Basel). The title list of the Latin 1538 edition states that the edition is *the first* (see the list of printed editions of the *Almagest* in Section 1): *Nunc primum edita, Interprete Georgio Trapezuntio*. In connection with this, the question arises: How reliably were the manuscripts dated on which the edition of 1528 (after Trapezuntius; no. 36 in our list in Section 1) and the edition of 1515 (now considered extremely rare; no. 35 in Section 1) were based? So far as we know, one more edition, of 1496, contains no star catalog. N. A. Morozov describes as follows the peculiarities that caused his interest to the problem of dating the *Almagest*: "I ... set to comparison of the latitudes given therein (in the Latin 1537 edition—*Authors*) with their modern state, recalculating for that into latitudes and longitudes the direct ascents and declinations from *Astronomischen Jahrbuch* of 1925. At the very first calculation, for Regulus, I was astonished: *I got the position relevant for the 16th century, the time of printing the book, but not for the 2nd century AD*. I took the Virgo's Spike and three other big stars, and again got the same thing: the longitudes Ptolemy gives are the ones of the 16th century! ... 'But how—I thought,—did Bode (whom I had not read in origin by the moment) and a number of other astronomers, like Bishop Montigno, derive the date in the 2nd century AD for this book?' ... The next morning ... I went to the Pulkov observatory in order to check these astonishing results by the first editions of the *Almagest* kept there ... I took from the shelf the first Greek edition (of 1538—*Authors*) and saw with surprise that *in it all longitudes are 20° (±10') less than in my Latin book, and the time of compilation is thus moved* by one and a half thousands years into the depth of centuries, if we count the longitudes given there from the spring equinoctial point . . . My bewilderment dispelled: Bode calculated from the Greek edition of 1538, and I from the precedent Latin edition of 1537. But instead, the question appeared: how strange it is that since the supposed time of Ptolemy to the time of the Greek edition, the precession amounted not to 15, 16, or 17 degrees, but to the round number 20°, and always with the same variation, plus or minus 10 minutes of arc" (Ref. 4, vol. 4, pp. 178–179).

The position of Bode is quite clear: why analyze the Latin "translation" of 1537 if there is an undoubtedly authentic Greek original of 1538 edition. N. A. Morozov expressed a suspicion that *the Latin text is primary, and the Greek is secondary* (but not vice versa, as is traditionally believed). Probably, the author of the 16th (or 15th) century who first published the ostensible translation, did not take care of the influence of precession, and when this was found out, introduced corrections in the Greek "original" text, moving it back to the 2nd century.

However, an objection against the primacy of the Latin text. In the 16th century, Ptolemy's book was published not as a historic document, but as a scientific treatise for the use of astronomers and students of astronomy. The data of the catalog, out of date because of precession, did not meet these purposes, so the translator could "replenish" the catalog by introducing the newest contemporary data (of the 16th century). As for the publisher of the Greek text (a year later), he could reason that, the Latin translation available, the text is no more needed as a textbook, and retrieve the original Ptolemy's numbers (attributing the catalog to the 2nd century). This argument is corroborated by the title page of the 1537 edition, where it is plainly written "ad hanc aetatem reducta, atque seorsum in studiorum gratium" ("to this time reduced specially for students"). Thus, this argument admits that the Latin edition is apocryphal (at least in what concerns the star catalog), but rejects that the Greek also is.

This argument is refuted by the fact that *in the Greek 1538 edition, all latitudes are systematically increased (improved), in comparison with the latitudes of the Latin 1537 edition by 25 minutes or are substituted by more precise values.* This is not a correction for precession, because latitudes are not subject to precession. The correction is circular, that is, the ecliptic is moved to the south, almost by the sun's diameter. As a result, the ecliptic acquired its true position, because the ecliptic plane now passes through the origin of coordinates; the ecliptic of the Latin edition was "bad" in the sense that the ecliptic plane did not pass through the origin of coordinates. Thus, the ecliptic of the Latin edition was not measured well, and in the later Greek edition the measurement was improved. Thus, we deal here with an *improvement* of the original Latin text. In the year that elapsed after the Latin edition, the observer made new observations, refined the measured stellar coordinates, and introduced the necessary corrections to the latitudes, thus making the ecliptic of the catalog closer to its true position.

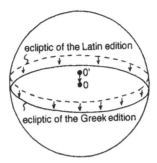

Figure A12. Positions of the ecliptics of the Greek and the Latin editions of the 16th century.

For a keen reader, we add the following explanatory comment. The ecliptic of the Greek edition is shown in Figure A12 as a dashed circumference, and the ecliptic of the Greek edition as a solid circumference. It is obvious that the "Latin ecliptic" does not pass through the center of the sphere. The "Greek ecliptic" is already in the correct position, because it is displaced down (parallel to the "Latin ecliptic") by 20 minutes of arc. Possibly, the error in the Latin edition is due to roughness of instruments or an insufficient accuracy of recalculation of equatorial coordinates into ecliptic.

Thus, "replenishing" Ptolemy's data in one respect (the account for precession), the editor of the Greek text improved them in the other. This is in no accord with the supposed primacy of the Greek text.

For an interested reader, we adduce the comparative table of "Greek" and "Latin" ecliptic latitudes of the stars of the first constellation in the *Almagest*, Ursa Minor (Table Ad. 3). The second column contains the latitudes from the Latin 1537 edition, the third the latitudes from the Greek 1538 editions and their variants from Refs. 22 and 17, and the last column contains differences: we subtract "Latin" latitudes from "Greek".

Thus, the difference between the "Latin" latitudes and the canonical latitudes in Refs. 22 and 17 is almost exactly equal to 25′ (for all stars of Ursa Minor).

4. Star maps of the Almagest.

The stars in the *Almagest* are described in relation to lines of constellations, which are assumed to be depicted in the sky. Using the catalog, an astronomer must first find the appropriate figure of a constellation, and then, looking into the catalog, find the star in the sky described, say, as "the star on the end of the tail" (the Polar star), or "the star above right knee" (a star in Ursa Major). If the astronomer has no star map with delineated constellations at hand, he will not find the star he is interested in. Of course, using the coordinates given in the catalog, he can try to fix the star with the help of instruments, but

Table Ad. 3

Numbers of stars in the *Almagest*	"Latin" latitudes	"Greek" latitudes	Differences
1 (α)	65°35′	66°00′	+25′
2 (δ)	69°35′	70°00′	+25′
3 (ε)	73°55′	74°20′	+25′
4 (ζ)	75°15′	75°20′ Ref. 22	+5′
		or	or
		75°40′ Ref. 17	+25′
5 (η)	77°15′	77°20′ Ref. 22	+5′
		or	or
		77°40′	+25′
6 (β)	72°25′	72°50′	+25′
7 (γ)	74°25′	74°50′	+25′
8 (A)	70°45′	71°10′	+25′

this means going through the process of measurement in the inverse order. Clearly, the catalog was designed for quickly finding stars in the sky, but not for this long process of "inverse measurement".

But then different astronomers must have absolutely identical star maps, in order to find exactly say, "the star above right knee". If the "knee" is drawn differently (or simply not quite accurately) in a map, then it is easy to mistake.

The exact finding of stars by the limbs of celestial animals, and transferring the positions from century to century and from country to country, without confusing names in the night sky, where neither legs nor tails are clearly seen, was only possible for the stars of the first and the second star magnitude. The stars of the third magnitude, of course, were already mistaken for each other, because one astronomer saw the ends of legs and tails of an imaginary animal lower or more to the right, while the other more to the left or higher.

In any case, an astronomer who compiled the catalog to an accuracy *within* 10′ (as the *Almagest* was claimed to be compiled) had to take into account the importance of the figures of constellation in various copies of a star map (which he could, say, disseminate among his disciples and colleagues) being *identical*.

As stated in the title page of the Latin edition, the *Almagest* was supplemented with 48 star maps, engraved by A. Dürer. Before the invention of printing, star maps only depicted the brightest stars, and the layout of the stars in the figure of a constellation varied from map to map. Only after the invention of engraving did the possibility of publishing detailed star maps arise, designed for use by many astronomers in various countries. Only a large-scale production of absolutely identical star maps could justify the great labor of creating a star map that showed the stars of the third and the fourth magnitude (as is done in the *Almagest*). Even if somebody set to this titanic work of creation of a single copy of such a map, it could not survive for centuries, because the unique copy of the map would soon moulder away, and a reproduction of it (accurate enough to make the copy usable) would mean carrying out all the work anew. The star maps of A. Dürer are the first really detailed maps of the starry sky.

It is obvious that the famous star maps of Dürer, engraved (as an inscription in Latin states) in 1515, got into the first Latin edition of the catalog of the *Almagest* (in 1537) many years after they disseminated among western astronomers as engravings. It is known from the history of technology that engraving as a technique for reproducing pictures first came into European use only in the beginning of the 15th century, and preceded immediately the invention of printing type. The technique first appeared in Holland and Flanders, and later came to France and Italy. The most ancient dated extant engraving is the woodcut *Saint Christopher*, dated 1423, 15 or 20 years before the invention of printing books (Ref. 4, vol. 4, pp. 221–222). That printed engravings were not known earlier, is obvious from the history of their invention. First, printings were manufactured in the same way as modern stamps are; in a piece of wood, recesses were cut with a stylus in the places intended to be white; applying an ink, the ready woodcut was pressed against paper, leaving a rough print. But this technique did not last long. In 1452, Tommaso Finiguera, a goldsmith from Florence, made the next natural step as he cut the picture in a silver plate, rubbed it with a mixture of oil and soot and pressed against a wet cloth, getting a fairly good print. Tommaso Finiguera repeated the process with sheets of wet paper, and saw that applying more paint to the engraving, it is possible to produce as many prints as desired. A further development of this technique was done by the famous Italian painter Mantegna (1431–1506; see Ref. 54, p. 756). He is the author of more than 20 engravings exposing mythologic, historical and religious scenes, for example, seven lists exposing a battle of marine gods (about 1470). Thus publishing engravings started and soon spread to Germany. Several years later, the name of Albrecht Dürer (1471–1528) became famous. He published in Nürnberg remarkable engravings, made in wood and in metal. The engravings were distinguished for the perfect shading, use of perspective and other merits. Soon, a school of prominent engravers arose.

Of course, printing star maps (dated by Dürer 1515) was easier than publishing the whole book with pictures (the *Almagest*). Dürer himself could make as many printings as he needed without any help of bookprinters. Apparently, he was not interested much in astronomy; at least, we do not know of any of his contributions to astronomy except the star maps.

Not an astronomer, Dürer, as he engraved star maps to order of an astronomer (or a publisher), made several gross errors in the maps, trying to keep the elegance of figures. We only expose here the most obvious examples.

In Dürer's map, Ara looks very beautiful and natural. But if we apply the map to the sky, it turns upside down, so that the tongue of its flame is directed downwards (see Figure A13). A natural question arises: What real observer could imagine it in this strange position?

Figure A13. The absurd position of Ara as depicted in Dürer's map (supplied to the first printed editions of the *Almagest*), in the real sky.

In Dürer's map, the constellation of winged Pegasus looks very beautiful and natural. Again, if we apply the map to the sky, we will see that "from dawn to sunset, Pegasus flies upside down, like a shot bird" (Ref. 4, vol. 4, p. 209; see Figure A14). It is also obvious that antique astronomers never meant to place Pegasus in this ridiculous position; this is a Dürer's blunder. Similarly, we see Hercules upside down as we apply the map to the sky.

All these (and some other) absurdities, obvious in the real sky, are not seen in the plane map. Therefore it is clear that their arrangement was chosen by Dürer according to the requirements of plane drawing; his errors are quite natural: he had before him a plane sheet of paper, not the real firmament, so he drew trying to create a piece of art. Production of the engravings, of course, took a great labor, so even if the absurdities made the astronomer, the author of the book, sick, he had nothing left to do but to print all this painting, more so because Dürer, who treated the maps as a piece of art only, could start to distribute copies himself, not waiting for the appearance of the book.

Dürer's upside down Pegasus worried, for example, Copernicus. Publishing his catalog, which, as we know, is a variant of the catalog of the *Almagest*, Copernicus made an attempt to "correct" the description of Pegasus. Not daring to redraw the star maps of Dürer, which he probably considered to be a reproduction of some lost ancient maps, he only altered the order of lines in the description of Pegasus, that is, made the first lines the last and vice versa. For example, in the *Almagest* "the star in the muzzle" has number 17

Figure A14. The absurd position of Pegasus as depicted in Dürer's map (supplied to the first printed editions of the *Almagest*), in the real sky.

(in Pegasus); Copernicus puts it first. Vice versa, the "star on the navel, which is common to the head of Andromeda" that goes first in the *Almagest*, is the last in Copernicus' catalog (number 20). But such a correction is quite naive and futile for the simple reason that it affected the table, but not the real sky, since the definition of stars in relation to the limbs of the figure remained the same. N. A. Morozov writes, "The Copernicus' attempt to correct the order of descriptions of the limb rather than the wrong figure itself is, of course, very naive, but it is a fact; he made alterations in the order of the *Almagest* in no other constellation" (Ref. 4, vol. 4, p. 225).

This is an evidence of a latent battle of the common sense of astronomers of the 16th century against the astronomic absurdity of some fragments of Dürer's star maps, hallowed by the Ptolemy's authority.

Admitting Dürer's authorship of all absurdities in the positions of constellations, we infer that any star map that reproduces Dürer's errors was made after Dürer. Let us now return to the *Almagest*. As we have noted many times, in the catalog of the *Almagest*, the positions of faint stars are described in verbal expressions of the type "in the muzzle of Pegasus", "above the left knee", "the more advanced of the two stars on the horn (of Aries)", etc. *It follows directly from the text that the maps the Almagest is supplemented with are intentional.* Indeed, let us again look at Pegasus. In the *Almagest*, the first star of the constellation is "the star on the navel" and "the star "in the muzzle" is the last (see Ref. 17, p. 358). The stars in the catalog are listed from north to south, so "the star in the muzzle" is to the north from "the star on the navel" (indeed, their latitudes in the *Almagest* are 26° and 22°30'). So, the author of the *Almagest* moves in the right direction, from the north to the south, and thus confirms the strange upside down position of Pegasus in the sky. A similar situation is with other constellations: the author means the maps supplemented to the *Almagest*.

But in this case, *the author of the catalog refers to the maps that contain Dürer's absurdities.* Therefore, all that verbal descriptions could not appear in the text of the *Almagest* before 1515. Thus, the conjecture arises that not only the star catalog, but also some (very essential!) chapters of the *Almagest* were created in their final form (or edited) as late as in the 16th century, right before its printing.

Each of the above items may be explained with more or less strain and ingenuity within the frames of traditional chronology. However, taken together they constitute too much proof that the main body of the *Almagest* belongs to the epoch of Renaissance.

N. A. Morozov concludes: "All this forces me to look at the *Almagest* as a summary of all astronomic knowledge and observations that accumulated since the establishment of zodiacal constellations in the first centuries of the Christian era till the 16th century; and some data it contains may and must belong to many preceding centuries. The goal of a serious researcher of this book is to determine which of the advances belong to which century" (Ref. 4, vol. 4, p. 218).

In vol. 4 of Ref. 4, an investigation of other star catalogs, in particular of Copernicus', Ulugh Beg's, As-Sûfi's, and Abū'l-Vafā's is carried out. We omit this material here, referring an interested reader to Ref. 4.

In conclusion, we formulate our hypothesis. Possibly, Hipparchus and Ptolemy are astronomers who really existed and were separated by approximately 200 to 300 years (as is presumed in the traditional chronology). However, the time of their lives should be displaced to the future by 1000 or 1300 years. Possibly, the activities of Hipparchus should be attributed to about the 10th–11th centuries, and Ptolemy's to the 13th–14th centuries. The *Almagest* was printed soon after it was finally written in the 14th–15th centuries.

In Figure A15 we reproduce the engraving of A. Dürer, a portrait of the famous medieval Roman emperor Maximilian Pius (1493–1519), in the reign of which particularly many manuscripts of the *Almagest* appeared (see the above chronological diagrams). Recall that tradition attributes the *Almagest* to the reign of Roman emperor Pius (who in his turn is traditionally attributed to the 2nd century AD). Figure A16 reproduces a medieval portrait of Ptolemy.

11. The *Almagest* and Halley's discovery of proper motions of stars

It is traditionally considered that the proper motions of stars were discovered in 1718 by E. Halley. Concerning this, P. G. Kulikovskiĭ tells: "E. Halley

ImpratorCæfarDiuusMaximilianus
Pius Frlix Auguſtus

Figure A15. Dürer's portrait of emperor Maximilian Pius Gabsburg, in whose reign the *Almagest* was first printed.

(1656–1742), as he compared contemporary positions of Arcturus, Sirius and Aldebaran with their position in Hipparchus' catalog, discovered proper motions of these stars: in 1850 years that elapsed (the date about the 2nd century BC is assumed for the Hipparchus' catalog; $1718 + 132 = 1850$ years.— *Authors*), the ecliptic latitudes of these stars altered by 60', 45' and 6'" (Ref. 55, p. 219). The first question that arises in connection with this is: How could he find the proper motion of Aldebaran? The fact is that the division value of Ptolemy's catalog (based on Hipparchus' catalog) is 10', while Aldebaran moved by mere 6'. Clearly, it is senseless to discuss an effect the magnitude of which is below the value of division of the measuring instrument. So how could Halley discover the proper motion of Aldebaran that had displaced by mere 6' for the ostensible 2000 years that had elapsed. Another question: What proper motions did Halley ascribe to Arcturus and Sirius? The same book communicates: "In 1738, G. Cassini (1677–1756) determined precisely the proper motion of Arcturus from comparison of the results of his

Figure A16. Medieval portrait of Ptolemy.

measurements with observations of J. Richet (1677–1756), made 60 years earlier" (Ref. 55, p. 219). Thus, Halley had determined the proper motion of Arcturus "inexactly". Still more inexactly could he determine the proper motion of Sirius, which moves slower than Arcturus.

So we naturally come to a somewhat unexpected question: Did Halley really discover proper motions of stars (even though in principle?).

It is relevant to remark here that Halley was not the first to raise the question whether stars move. The question was actively discussed by astronomers in the 15th–16th centuries, long before Halley. Furthermore, within the frames of traditional chronology, the question had been raised by antique astronomers, ostensibly approximately two thousands years before Halley. It turns out that the question whether the stars move had been considered by Hipparchus. The famous Roman historian Pliny (traditionally attributed to 23–79 AD) wrote "This Hipparchus ... studied a new star that appeared in his time; that it moved as it shone (probably, a comet is meant. — *Authors*) gave an idea to him that those (luminaries) which we think to be fixed may often alter and move; so he set about an affair, bold even for a god, to enumerate stars for the posterity, and count the luminaries, inventing instruments to determine the places and brightness of separate stars, in order to make it easily possible to see, whether they disappear or appear again, and whether they move, or increase or decrease (in brightness), leaving the sky as a legacy for the posterity, provided there were anybody to accept this legacy" (Pliné, l'Ancient. *Histoire Naturelle, Livre II.P.*, Colléction Budé, 1950, XXI⁺ 282 p.; see also Ref. 16, p. 31).

It is traditionally believed that this important question was considered also by Ptolemy. Having studied this question, which was of fundamental importance for him, he came to the conclusion that the stars are fixed. Thus, the statement of the question cannot be ascribed to Halley.

Why did nobody before Halley compare contemporary positions of stars with the positions given in the *Almagest*, in order to reveal the proper motion? The fact is that the very idea of such a comparison goes back to Ptolemy, so it was no news for medieval astronomers. Attempts of this kind would be quite natural, and could lead to the conclusion that the stars move; say, sharp outlies in Ptolemy's measurement could be taken for the proof of the existence of proper motion. The astronomers who lived in the beginning of the 17th century (a hundred years before Halley) might have determined the proper motions just as well as Halley, using the catalog of Tycho Brahe (it has accuracy of 1 minute of arc, and is dated 1582–1588) for comparison with the Ptolemy's catalog.

Let us try to understand the point of view of medieval astronomers of the 16th–17th centuries (a hundred years before Halley). A priori, two views on Ptolemy's *Almagest* are possible.

1) Let us first assume the astronomers already kept to the chronology of Scaliger and Petavius, according to which the reign of Antoninus Pius (to which the observations of stars are associated in the *Almagest*) began in 138 AD. In this case, they should try to discover the proper motions of stars, using this ostensibly ancient catalog (about fifteen centuries old). In this comparison, they could use, say, Arcturus, the brightest star in the northern sky. However, no such attempts were ever fixed by the traditional history of astronomy of the 15th–17th centuries, although such an attempt would lead to the conclusion that at least Arcturus does move.

2) Let us now assume that the astronomers of the 16th–17th centuries regarded the *Almagest* as a comparatively new document (say, of the 13th–15th centuries), or as a document whose date of creation is unknown. In this case, their attitude to the *Almagest* would be different. If the astronomers thought that the document had been created not very long before, then the small interval of time that had elapsed since the time of compilation could be treated as too small for detecting proper motions of stars. Furthermore, if this catalog was treated as medieval, then the inaccuracy of the *Almagest*, well known to professional astronomers (as well as the roughness of the division value in the catalog) merely aborted a possibility of such a comparison for

individual stars. Also, a calculation would be impossible in the case that the date of compilation was not known.

Since the history of astronomy tells nothing about attempts of astronomers of the 16th–17th centuries to detect proper motions of stars using the *Almagest*, we may state the conjecture that the astronomers did not regard the *Almagest* as being sufficiently old (and sufficiently accurate).

Thus, a serious scientist of the 16th–17th century, who treated the *Almagest* as a medieval document had to come to the conclusion that the accuracy of coordinates in the *Almagest* does not provide a possibility for detection of proper motions of stars. On the other hand, if the *Almagest* had been treated as an ancient text (created, say, about the 2nd century AD), then the very idea of using it for the detection of proper motion would be so obvious (and the question about the proper motion was considered as so important) that it appears to be absolutely improbable that this idea would have never occurred to anybody much before Halley.

Let us now try to explain why in the times of Halley, it was already possible to detect the proper motion of some stars (say, of Arcturus and Sirius), but it was still impossible to determine more or less accurately the velocities (or the displacements).

The first precise catalog (with the accuracy 1′) was the catalog of Tycho Brahe, compiled in 1582–1588. The displacements of Arcturus and Sirius in the 100 years that had elapsed by the time of Halley amounted to approximately 3′ and a little more than 2′ respectively. Having the availability of a precise catalog of stellar positions in the beginning of the 18th century, one already could suspect the proper motion of Arcturus and Sirius, although the accuracy of Tycho Brahe's catalog was still too low to find the velocities of motion of these two stars. It turns out that the requisite catalog for the beginning of the 18th century existed: this was the catalog of John Flamsteed, which Halley in fact used before it was published (borrowing a version of the catalog from I. Newton, who at the time was interested in chronology).

Thus, our opinion is that Halley compared the catalogs of Flamsteed and of Tycho Brahe and derived a conclusion about the existence of proper motion of Arcturus, Sirius and Aldebaran.

If so, his indication of the Aldebaran's proper motion acquires a natural explanation. The fact is that Halley used a version of Flamsteed's catalog that was not final and was not refined, which could indicate the position of Aldebaran with an error. Flamsteed himself at the moment considered the catalog not ready for publication; some evidence exists that Flamsteed still refined the position of Aldebaran (Flamsteed's letter to A. Sharp of September 13, 1718; see *A. Baily, An Account of the Life of Sir John Flamsteed* (London, 1835)).

But why did Halley refer to the *Almagest*, not to the catalog of Tycho Brahe? Possibly, in Halley's time, the date for the *Almagest* "calculated" by Scaliger and Petavius (138 AD) had already been canonized. Halley could refer to the *Almagest* to add ponderability to his discovery, for the simple reason that in this case the detected displacements of stars looked more impressive. Calculating the displacement of Arcturus from the catalog of Tycho Brahe, Halley only obtained 3' (in comparison with the nominal accuracy of the catalog amounting to 1'), while a calculation of the displacement from the Ptolemy's *Almagest* (purportedly, of 138 AD) gave a much more notable displacement, about 1 degree of arc. Apparently, Halley compared this displacement with the nominal accuracy of the catalog (10'), ignoring the question of the true accuracy.

This consideration once more brings about the conclusion that in the 16th century, the *Almagest* probably was not treated as an ancient (aged fifteen centuries) document, but in the beginning of the 18th century, after the chronological works of Scaliger and Petavius, its being ancient had already been admitted (and canonized).

Appendix

Table Ap. 1.

	No. in Ref. 21	α_{1900}	δ_{1900}	v_α	v_δ	Magnitude
	6	$0^h\ 1^m\ 8^s.0$	$-49°38'$	560	-37	5.77
11 β Cas	21	$0^h\ 3^m\ 50^s.0$	$58°36'$	527	-178	2.42
	77	$0^h\ 14^m\ 52^s.0$	$-65°28'$	1708	1163	4.34
	98	$0^h\ 20^m\ 30^s.0$	$-77°49'$	2223	326	2.90
	159	$0^h\ 32^m\ 12^s.0$	$-25°19'$	1383	-8	5.71
	173	$0^h\ 35^m\ 31^s.0$	$-24°21'$	640	-329	6.24
	176	$0^h\ 35^m\ 44^s.0$	$-60°\ 1'$	886	451	5.79
24 η Cas	219	$0^h\ 43^m\ 3^s.0$	$57°17'$	1101	-523	3.64
	222	$0^h\ 43^m\ 8^s.0$	$4°46'$	752	-1142	5.82
μ Cas	321	$1^h\ 1^m\ 37^s.0$	$54°26'$	3430	-1575	5.26
52 τ Cet	509	$1^h\ 39^m\ 25^s.0$	$-16°28'$	-1718	860	3.65
	637	$2^h\ 6^m\ 19^s.0$	$-51°19'$	2108	651	6.28
	660	$2^h\ 10^m\ 57^s.0$	$33°46'$	1155	-240	5.07
	753	$2^h\ 30^m\ 36^s.0$	$6°25'$	1807	1459	5.92
18 ι Per	937	$3^h\ 1^m\ 51^s.0$	$49°14'$	1267	-81	4.17
	1006	$3^h\ 15^m\ 36^s.0$	$-62°57'$	1332	659	5.48
	1008	$3^h\ 15^m\ 56^s.0$	$-43°27'$	3056	744	4.30
	1010	$3^h\ 16^m\ 2^s.0$	$-62°53'$	1328	655	5.16
23 δ Eri	1136	$3^h\ 38^m\ 27^s.0$	$-10°\ 6'$	-92	744	3.72

Table Ap. 1. (Continuation)

	No. in Ref. 21	α_{1900}	δ_{1900}	v_α	v_δ	Magnitude
40 o^2 Eri	1325	$4^h\ 10^m\ 40^s.0$	$-7°49'$	-2225	-3418	4.48
	1614	$4^h\ 55^m\ 51^s.0$	$-5°52'$	557	-1089	6.50
15 λ Aur	1729	$5^h\ 12^m\ 6^s.0$	$40°\ 1'$	528	-659	4.85
	2083	$5^h\ 51^m\ 44^s.0$	$-50°24'$	74	568	5.0
	2102	$5^h\ 53^m\ 20^s.0$	$-63°\ 7'$	135	540	4.53
9 α C Ma	2491	$6^h\ 40^m\ 44^s.6$	$-16°35'$	-545	-1211	-1.60
10 α C Mi	2943	$7^h\ 34^m\ 4^s.0$	$5°29'$	-706	-1032	0.48
78 β Gem	2990	$7^h\ 39^m\ 12^s.0$	$28°16'$	-623	-52	1.21
	2998	$7^h\ 39^m\ 51^s.0$	$-44°55'$	-72	-563	5.22
	3018	$7^h\ 41^m\ 51^s.0$	$-39°59'$	-293	1663	5.39
	3384	$8^h\ 28^m\ 57^s.0$	$-31°11'$	-1119	757	6.36
	3951	$9^h\ 55^m\ 15^s.0$	$32°25'$	-522	-436	5.60
	4098	$10^h\ 21^m\ 54^s.0$	$49°19'$	81	-892	6.50
53 ξ U Ma	4375	$11^h\ 12^m\ 51^s.0$	$32°\ 6'$	-431	-593	4.41
	4414	$11^h\ 21^m\ 42^s.0$	$3°33'$	-722	177	6.19
	1414	$11^h\ 21^m\ 43^s.0$	$3°33'$	-723	169	7.90
	4486	$11^h\ 33^m\ 29^s.0$	$45°40'$	-594	18	6.39
	4523	$11^h\ 41^m\ 45^s.0$	$-39°57'$	-1538	393	5.04
	4540	$11^h\ 45^m\ 29^s.0$	$2°20'$	742	-277	3.80
5 β Vir	4550	$11^h\ 47^m\ 13^s.0$	$38°26'$	3994	-5800	6.46
	4657	$12^h\ 10^m\ 2^s.0$	$-9°44'$	31	-1024	6.12
	4710	$12^h\ 17^m\ 51^s.0$	$-67°\ 5'$	-748	243	6.38
43 β Coma	4983	$13^h\ 7^m\ 12^s.0$	$28°23'$	-799	876	4.32
	5019	$13^h\ 13^m\ 10^s.0$	$-17°45'$	-1075	-1076	4.80
	5072	$13^h\ 23^m\ 32^s.0$	$14°19'$	-237	-583	5.16
	5183	$13^h\ 42^m\ 0^s.0$	$6°51'$	-513	-114	6.32
	5189	$13^h\ 43^m\ 10^s.0$	$-35°12'$	-522	-178	6.47
	5209	$13^h\ 45^m\ 50^s.0$	$-23°53'$	-575	-310	6.48
5 θ Cen	5288	$14^h\ 0^m\ 48^s.0$	$-35°53'$	-521	-522	2.26
16 α Boo	5340	$14^h\ 11^m\ 6^s.0$	$19°42'$	-1098	-2003	0.24
	5455	$14^h\ 31^m\ 41^s.0$	$-11°53'$	-876	359	6.24
α Cen	5460	$14^h\ 32^m\ 48^s.0$	$-60°25'$	-3606	705	0.33
	5568	$14^h\ 51^m\ 37^s.0$	$-20°58'$	1041	-1745	5.76
v^2 Lup	5699	$15^h\ 15^m\ 3^s.0$	$-47°57'$	-1621	-275	5.71
41 γ Ser	5933	$15^h\ 21^m\ 50^s.0$	$15°59'$	307	-1292	3.86
15 ρ Cor B	5968	$15^h\ 57^m\ 13^s.0$	$33°36'$	-200	-774	5.43
	6014	$16^h\ 4^m\ 16^s.0$	$6°40'$	235	-744	6.02
	6060	$16^h\ 10^m\ 11^s.0$	$-8°\ 6'$	227	-508	5.56
26 ε Sco	6241	$16^h\ 43^m\ 41^s.0$	$-34°\ 7'$	-613	-256	2.36
	6401	$17^h\ 9^m\ 12^s.0$	$-26°27'$	-464	-1146	5.33
	6401	$17^h\ 9^m\ 12^s.0$	$-26°27'$	-497	-1137	5.29
	6416	$17^h\ 11^m\ 28^s.0$	$-46°32'$	975	213	5.58
	6426	$17^h\ 12^m\ 9^s.0$	$-34°53'$	1167	-176	5.89
	6458	$17^h\ 16^m\ 55^s.0$	$32°36'$	126	-1047	5.36
	6518	$17^h\ 25^m\ 18^s.0$	$67°23'$	-529	0	6.31

Table Ap. 1. (Continuation)

	No. in Ref. 21	α_{1900}	δ_{1900}	v_α	v_δ	Magnitude
	6573	$17^h\ 33^m\ 57^s.0$	$61°57'$	253	-513	5.31
86 μ Herc	6623	$17^h\ 42^m\ 33^s.0$	$27°47'$	-313	-748	3.48
	6752	$18^h\ 0^m\ 24^s.0$	$2°31'$	256	-1097	4.07
58 η Ser	6869	$18^h\ 16^m\ 8^s.0$	$-2°52'$	-554	-697	3.26
44 χ Dra	6927	$18^h\ 22^m\ 51^s.5$	$72°41'$	521	-356	3.57
	7373	$19^h\ 20^m\ 12^s.1$	$11°44'$	722	640	5.16
	7644	$19^h\ 55^m\ 32^s.3$	$-67°35'$	845	-680	6.07
	7703	$20^h\ 4^m\ 37^s.7$	$-36°21'$	449	-1568	5.32
	7722	$20^h\ 9^m\ 3^s.1$	$-27°20'$	1244	-178	5.73
	7875	$20^h\ 31^m\ 45^s.7$	$-50°53'$	309	-569	5.12
3 η Cep	7957	$20^h\ 43^m\ 15^s.3$	$61°27'$	91	822	3.43
	8085	$21^h\ 2^m\ 24^s.8$	$38°15'$	4135	3250	5.21
	8086	$21^h\ 2^m\ 26^s.3$	$38°15'$	4122	3112	6.03
	8148	$21^h\ 13^m\ 58^s.8$	$-26°46'$	-539	-352	6.56
	8387	$21^h\ 55^m\ 42^s.6$	$-57°12'$	3940	-2555	4.59
	8697	$22^h\ 47^m\ 19^s.9$	$9°18'$	522	49	5.16
53 β Peg	8775	$22^h\ 58^m\ 56^s.0$	$27°32'$	188	139	2.60
	8829	$23^h\ 7^m\ 56^s.6$	$-63°14'$	477	-422	6.12
	8832	$23^h\ 8^m\ 27^s.8$	$56°37'$	2073	299	5.56
17 ι Psc	8969	$23^h\ 34^m\ 48^s.3$	$5°\ 5'$	375	-432	4.13

Table Ap. 2. Named Stars.

	Modern data (by Ref. 21)				Almagest		
The name	α_{1900}	δ_{1900}	v_α	v_δ (0.0001″/year)	Baily's number	l	b
Alcyone (η Tau)	$3^h\ 41^m.54$	$23°\ 47'.77$	19	-44	410	$32°\ 30'$	$3°\ 40'$
Aldebaran (α Tau)	$4^h\ 30^m.18$	$16°\ 18'.50$	65	-189	393	$42°\ 40'$	$-5°\ 10'$
Alderamin (α Ceph)	$21^h\ 16^m.18$	$62°\ 9'.72$	150	52	78	$346°\ 40'$	$69°\ 0'$
Algeiba (γ Leo)	$10^h\ 14^m.46$	$20°\ 20'.85$	307	-151	467	$122°\ 10'$	$8°\ 30'$
Algenib (γ Peg)	$0^h\ 8^m.09$	$14°\ 37'.67$	3	-7	316	$342°\ 10'$	$12°\ 30'$
Alhena (γ Gem)	$6^h\ 31^m.93$	$16°\ 29'.08$	43	-44	440	$72°\ 0'$	$-7°\ 30'$
Alioth (ε U Ma)	$12^h\ 49^m.63$	$56°\ 30'.15$	109	-10	33	$132°\ 10'$	$53°\ 30'$
Alkaid (η U Ma)	$13^h\ 43^m.60$	$49°\ 48'.75$	-126	-14	35	$149°\ 50'$	$54°\ 0'$

Table Ap. 2. Named Stars (Continuation).

The name	Modern data (by Ref. 21)				Almagest		
	α_{1900}	δ_{1900}	v_α	v_δ	Baily's	l	b
			(0.0001''/year)		number		
Almaak (γ And)	$1^h 57^m.76$	$41°\,51''.00$	46	−48	349	$16°\,50'$	$28°\,\,0'$
Alnilam (ε Ori)	$5^h 31^m.14$	$-1°\,15''.95$	−3	−2	760	$57°\,20'$	$-24°\,50'$
Alnitak (γ Ori)	$5^h 35^m.71$	$-1°\,59''.72$	−1	−2	761	$58°\,10'$	$-25°\,40'$
Alphard (α Hyd)	$9^h 22^m.67$	$-8°\,13''.50$	−18	28	905	$120°\,\,0'$	$-23°\,\,0'$
Alphekka (α Cor B)	$15^h 30^m.49$	$27°\,\,3''.07$	120	−91	111	$194°\,40'$	$44°\,40'$
Alpheratz (α And)	$0^h\,\,3^m.22$	$28°\,32''.30$	137	−158	315	$347°\,50'$	$26°\,\,0'$
Alshain (β Aql)	$19^h 50^m.40$	$6°\,\,9''.42$	−43	−479	287	$274°\,50'$	$27°\,10'$
Altair (α Aql)	$19^h 45^m.90$	$8°\,36''.25$	537	387	288	$273°\,50'$	$29°\,10'$
Antares (α Sco)	$16^h 23^m.27$	$-26°\,12''.60$	−7	−23	553	$222°\,40'$	$-4°\,\,0'$
Arcturus (α Boo)	$14^h 11^m.10$	$19°\,42''.18$	−1098	−1999	110	$177°\,\,0'$	$31°\,30'$
Arheb (α Lep)	$5^h 28^m.32$	$-17°\,53''.63$	−6	1	812	$55°\,50'$	$-41°\,30'$
Bellatrix (γ Ori)	$5^h 19^m.77$	$6°\,15''.55$	−12	−14	736	$54°\,\,0'$	$-17°\,30'$
Betelgeuse (α Ori)	$5^h 49^m.46$	$7°\,23''.32$	25	10	735	$62°\,\,0'$	$-17°\,\,0'$
Canopus (α Car)	$6^h 21^m.72$	$-52°\,38''.45$	26	22	892	$77°\,10'$	$-75°\,\,0'$
Capella (α Aur)	$5^h\,\,9^m.30$	$45°\,54''.00$	83	−427	222	$55°\,\,0'$	$22°\,30'$
Caph (β Cas)	$0^h\,\,3^m.80$	$58°\,36''.00$	527	−178	189	$7°\,50'$	$51°\,40'$
Castor (α Gem)	$7^h 28^m.22$	$32°\,\,6''.45$	−170	−102	424	$83°\,20'$	$9°\,40'$
Cor Caroli (α C Vn)	$12^h 51^m.35$	$38°\,51''.50$	−236	52	36	$147°\,50'$	$39°\,45'$
Deneb (α Cygn)	$20^h 38^m.00$	$44°\,55''.00$	0	0	163	$309°\,10'$	$61°\,\,0'$
Denebola (β Leo)	$11^h 43^m.96$	$15°\,\,7''.53$	−497	−119	488	$144°\,30'$	$11°\,50'$
Denedkaitos (β Cet)	$0^h 38^m.57$	$-18°\,32''.13$	232	36	733	$335°\,40'$	$-20°\,20'$
Dubhe (α U Ma)	$10^h 57^m.56$	$62°\,17''.45$	−118	−71	24	$107°\,40'$	$49°\,\,0'$

Table Ap. 2. Named Stars (Continuation).

The name	Modern data (by Ref. 21)				Almagest		
	α_{1900}	δ_{1900}	v_α	v_δ	Baily's		
			(0.0001″/year)		number	l	b
Fomalhaut (α Ps A)	$22^h 52^m.13$	$-30°\ 9'.13$	336	-161	670	307° 0′	$-20°\ 20'$
Hadar (β Cen)	$13^h 56^m.76$	$-59°\ 53'.43$	-20	-23	970	204° 10′	$-45°\ 20'$
Hamal (α Ar)	$2^h\ 1^m.53$	$22°\ 59'.38$	190	-144	375	10° 40′	10° 0′
Izar (ε Boo)	$14^h 40^m.62$	$27°\ 29'.75$	-51	18	103	180° 0′	40° 15′
Markab (α Peg)	$22^h 59^m.78$	$14°\ 40'.03$	62	-38	318	326° 40′	19° 40′
Menkar (α Cet)	$2^h 57^m.05$	$3°\ 41'.85$	-12	-74	713	17° 40′	$-12°\ 20'$
Mirfak (α Per)	$3^h 17^m.18$	$49°\ 30'.32$	25	-22	197	34° 50′	30° 0′
Polaris (α U Ma)	$1^h 22^m.56$	$88°\ 46'.43$	46	-4	1	60° 10′	66° 0′
Procyon (α C Mi)	$7^h 34^m.07$	$5°\ 28'.88$	-706	-1029	848	89° 10′	$-16°\ 10'$
Pas Alh (α Orph)	$17^h 30^m.29$	$12°\ 37'.97$	117	-227	234	234° 50′	36° 0′
Regulus (α Leo)	$10^h\ 3^m.05$	$12°\ 27'.37$	-249	-3	469	122° 30′	0° 10′
Spica (α Vir)	$13^h 19^m.92$	$-10°\ 38'.37$	-43	-33	510	176° 40′	$-2°\ 0'$
Lyra (Vega) (α Lyr)	$18^h 33^m.55$	$38°\ 41'.43$	200	285	149	257° 20′	62° 0′
Rigel (β Ori)	$5^h\ 9^m.43$	$-8°\ 19'.02$	-3	-2	768	49° 50′	$-31°\ 30'$
Sirius (α C Ma)	$6^h 40^m.74$	$-16°\ 34'.73$	-545	-1211	818	77° 40′	$-39°\ 10'$

References

1. Newton, R. R., *The Crime of Claudius Ptolemy*, The Johns Hopkins University Press, Baltimore, MD, 1977.
2. Berry, A., *Short Story of Astronomy*, Dover Publications, New York, 1961/
3. Pannekoek, A., *A History of Astronomy*, Interscience Publishers, New York, 1961.
4. Morozov, A. N., *Khristos (History of Science in the Light of Natural Science)*, vols. 1–7, GIZ, Moscow - Leningrad, 1928–1932 (Russian).
5. Fomenko, A. T., On some properties of the second derivative of the moon's elongation and related statistical regularities, *Voprosy Vychislitel'noi i Prikladnoi Matematiki* **63**, Akad. Nauk Uzb. SSR, Tashkent, 1981, pp. 136–150 (Russian).
6. Fomenko, A. T., The jump of the second derivative of the moon's elongation, *Celestial Mechanics*, **29** (1981), 3–40.
7. Fomenko, A. T., Computation of the second derivative of the moon's elongation and statistical regularities in distribution of certain astronomic data, *Issledovanie operatsiĭ i ASU* **20**, Kiev Univ. Publ., Kiev, 1983, pp. 89–113 (Russian).
8. Fomenko, A. T., New empirico-statistical method for ordering texts with applications to dating problems, *Doklady Akad. Nauk SSSR* **268** no. 6 (1983), 1322–1327 (Russian).
9. Fomenko, A. T., A method for recognition of duplicates and some applications, *Doklady Akad. Nauk SSSR* **258** no. 6 (1981), 1326–1330 (Russian).
10. Fomenko, A. T., New empirico-statistical dating methods and statistics of certain astronomical data, *The First World Congress of the Bernoulli Society for Mathematical Statistics and Probability Theory, Abstracts*, vol. 2, Nauka, Moscow, 1986, p. 892.
11. Fomenko, A. T., New empirico-statistical method for detecting parallelisms and dating duplicates, *Problemy Ustoičivosti Stokhastičeskikh Modelei. Trudy Seminara*. VNIISI, Moscow, 1984, pp. 154–177 (Russian).
12. Fomenko, A. T., Recognition of relations and stratified structures in narrative texts, *Problemy Ustoičivosti Stokhastičeskikh Modelei. Trudy Seminara*, VNIISI, Moscow, 1987, pp. 15–28 (Russian).

13. Fomenko, A. T., *Some New Empirico-Statistical Methods of dating and the Analysis of Present Global Chronology*, 1981, The British Library, Dept. of Printed Books Cup 918/87.

14. Kalashnikov, V. V., Nosovsky, G. V., and Fomenko, A. T., Geometry of moving configurations of stars and dating the *Almagest*, *Problemy Ustoičivosti Stokhastičeskikh Modelei. Trudy Seminara*, VNIISI, Moscow, 1988, pp. 59–78 (Russian).

15. Kalashnikov, V. V, Nosovsky, G. V., and Fomenko, A. T., A date for the *Almagest* from varying star configurations, *Doklady Akad. Nauk SSSR* **307** no. 4 (1989), 828–832 (Russian).

16. Bronšten, V. A., *Klavdii Ptolemei*, Nauka, Moscow, 1988 (Russian).

17. *Ptolemy's Almagest*, Translated and annotated by G. J. Toomer, Springer, London, 1984.

18. Klimišin, I. A., *Otkrytie Vselennoi*, Nauka, Moscow, 1987 (Russian).

19. Newcomb, S., Tables of the Motion of the Earth on its Axis and around the Sun, Part I, *Astronomical Papers*, vol. VI, 1898.

20. Fomenko, A. T., *Simplektičeskaia Geometriia i topologiia. Metody i Prilozeniya*, Moscow Univ. Publ., Moscow, 1988 (Russian).

21. Hofflit, D., *The Bright Star Catalogue*, Yale Univ. Obs., New Haven, CT, 1982.

22. Peters, C. H. F. and Knobel, E. B., *Ptolemy's Catalogue of Stars. A Revision of the Almagest*, The Carnegie Institute of Washington, Washington, DC, 1915.

23. Kinoshita, H., *Formulas for Precession*, 02138, Feb. 28, Smithsonian Inst. Astrophys. Observatory, Cambridge, MA, 1975.

24. Ptolemy, *The Almagest*, translated by R. Catesby Taliaferro, Great Books of the Western World, vol. 16, The University of Chicago, Encyclopaedia Britannica, Chicago, IL, 1952.

25. Copernici, Nicolai, *Revolutionibus Orbium Caelestium. Lib. VI*, edited by G. Ioachimi, Thoruni, 1873.

26. Bode, J. E., *Claudius Ptolemaeus Beobachtung und Beschreibung der Gestirne*, Berlin, 1795.

27. Efremov, Yu. N. and Pavlovskaia, E. D., Dating the *Almagest* from proper motions of stars, *Doklady Akad. Nauk SSSR* **294** no. 2 (1987), 310–313 (Russian).

28. Baily, F., The Catalogues of Ptolemy, Ulugh Beg, Tycho Brahe, Halley and Hevelius, Deduced from the Best Authorities, *Royal Astr. Soc. Memoirs*, vol. XIII, London, 1843, pp. 1–248.

29. Fricke, W. and Kopf, A., *FK4*, Veröf. Astr. Inst. Heidelberg, Heidelberg, 1975.

30. Efremov, Yu. N. and Pavlovskaia, E. D., Determination of the epoch of observations for the star catalog of the *Almagest* from proper motions of stars, *Istoriko-astronomicheskie Issledovaniya*, Nauka, Moscow, 1989, pp. 175–192 (Russian).

31. Fomenko, A. T., *Metody Statističeskogo Analiza Narrativnykh Tekstov i Priloženiya k Khronologii*, Moscow State Univ. Publishers, Moscow, 1990.

32. Blažko, S. N., *Kurs Prakticheskoi Astronomii*, Nauka, Moscow, 1979 (Russian).

33. Ševčenko, M. Yu., Star catalog of Claudius Ptolemy: specific features of antique astrometric observations, *Istoriko-astronomicheskie Issledovaniya*, Nauka, Moscow, 1989, pp. 167–186 (Russian).

34. Tychonis Brahei, *Equitis dani Astronomorum Coryphaei vita Authore Petro Gassendo*, Regio ex Typographia Adriani Vlaco M.DC.LV, 1655.

35. Fomenko, A. T., Kalashnikov, V. V., Nosovsky, G. V., When was Ptolemy's star catalogue compiled in reality? Statistical analysis, *Acta Applicandae Mathematicae* **17** (1989), 203–229.

36. Fomenko, A. T., Kalashnikov, V. V., and Nosovsky, G. V., Statistical analysis and dating of the observations on which Ptolemy's star catalog is based, *Probability Theory and Mathematical Statistics. Proc. Fifth Vilnius Conference 1990 Moklas, Vilnius, Lithuania*, vol. 1, VSP, Utrecht, 1990, pp. 360–374.

37. Newton, R. R., *The Origins of Ptolemy's Astronomical Tables*, The Johns Hopkins University Press, Baltimore, MD, 1985.

38. Duboshin, G. N., *Spravočnik po Nebesnoi Mekhanike i Astrodinamike*, Nauka, Moscow, 1976 (Russian).

39. Abalakin, V. K., *Osnovnye Poniatiya Astronomii Efemerid*, Nauka, Moscow, 1979 (Russian).

40. Žitomirskiĭ, S. V., Antique views of the dimensions of the world, *Istoriko-astronomicheskie Issledovaniya*, Nauka, Moscow, 1983, pp. 291–326 (Russian).

41. Oppolzer, Th., *Kanon der Sonnen und Mondfinsternisse*, K. K. Hof- und Staatsdruckerei, 1887.

42. *Legenda o doctore Fauste*, Literaturnye Pamyatniki, Nauka, Moscow, 1978 (Russian).

43. *Rukopisnaia i Pečatnaia Kniga*, Nauka, Moscow, 1975 (Russian).

44. *Mahabharata* (B. L. Smirnov ed.), vol. 1–8, Akad. Nauk Turkmen SSR, Ashkhabad, 1955–1971.

45. Nikonov, V. A., *Imia i Obščestvo*, Nauka, Moscow, 1974.

46. Veselovskii, I. N., Aristarchus of Samos, Copernicus of the antique world, *Istoriko-astronomicheskie Issledovaniya*, Nauka, Moscow, 1961, p. 44 (Russian).

47. Cicero, *Dialogi (Dialogues)*, Nauka, Moscow, 1966 (Russian translation).

48. Cicero, *Filosofskie Traktaty (Philosophic Treatises)*, Nauka, Moscow, 1985 (Russian translation).

49. Titus Livius, *Rimskaia Istoria ot Osnovaniya Goroda (Roman History from Foundation of the City)*, vols. 1–6, Typ. E. Gerbek, 1897–1899 (Russian translation).

50. Ginzel, F. K., *Specieller Canon der Sonner- und Mondfinsternisse*, Berlin, 1899.

51. *Istoriko-matematičeskie Issledovaniya* 1, Fizmatgiz, Moscow, 1948.

52. Diophantus, *Arifmetika (Arithmetics)*, Fizmatgiz, Moscow, 1974 (Russian translation).

53. Archimedes, *Sočineniya (Works)*, Fizmatgiz, Moscow, 1962 (Russian translation).

54. *Sovietskiĭ Entsyklopedičeskiĭ Slovar (Soviet Encyclopedic Dictionary)*, Sovietskaia Entsyclopediia, Moscow, 1979.

55. Kulikovskiĭ, P. G., *Zviozdnaia Astronomiya*, Nauka, Moscow, 1978.

56. *An Islamic Book of Constellations*, Bodleian Library. Oxford. 1965, Oxford Univ. Press.

57. H. C. F. C. Schjellerup, *Description des étoiles fixes composée au milieu du dixième siècles de notre ère, par l'astronome persan Abd-al-Rahman Al-Sufi*, St. Petersbourg, 1874. (This contains a full translation of al-Ṣūfi's text).

58. Joseph M. Upton, A manuscript of "The book of fixed stars" by Abd al-Rahman as-Sufi, *Metropolitan Museum Studies*, iv (1933), 179–197.

59. Fomenko A. T., Kalashnikov V. V., Nosovsky G. V., The dating of Ptolemy's *Almagest* based on the coverings of the stars and on lunar eclipses, *Acta Applicandae Mathematicae*. 1992, vol. 29, pp. 281–298.

60. Fomenko A. T., Mathematical statistics and problems of ancient chronology. A new approach. *Acta Applicandae Mathematicae*. 1989, vol. 17, pp. 231–256.

Index

'f' denotes figures, 't' denotes tables.